P9-CCM-715

30lbs: per sq: in
MODEL ENGINE CONSTRUCTION
J. ALEXANDER
Lost Technology Series
Reprinted by Lindsay Publications Inc.

Model Engine Construction

J. Alexander

Copyright 1986 by Lindsay Publications, Inc., Bradley, IL 60915. Originally published in 1894 by Whittaker and Co., 66, Fifth Avenue, New York.

All rights reserved. No part of this book may be reproduced in any form or by any means without written permission from the publisher.

ISBN 0-917914-44-9

4 5 6 7 8 9 0

MODEL ENGINE CONSTRUCTION

WITH PRACTICAL INSTRUCTIONS

TO ARTIFICERS AND AMATEURS.

BY

J. ALEXANDER.

CONTAINING

NUMEROUS ILLUSTRATIONS AND TWENTY-ONE WORKING DRAWINGS FROM ORIGINAL DRAWINGS BY THE AUTHOR, AND RE-DRAWN BY C. E. JONES.

WARNING

Remember that the materials and methods described here are from another era. Workers were less safety conscious then, and some methods may be downright dangerous. Be careful! Use good solid judgement in your work, and think ahead. Lindsay Publications, Inc. has not tested these methods and materials and does not endorse them. Our job is merely to pass along to you information from another era. Safety is your responsibility.

Write for a catalog or other unusual books available from:

Lindsay Publications, Inc.
PO Box 12
Bradley, IL 60915-0012

PREFACE.

Having had some experience in the construction of model steam-engines, I publish this work for the purpose of giving a few practical instructions to Artificers, but more especially to Amateurs who wish to make a working model steam-engine, provided they be supplied with the proper tools, and only know how to set about it. I commence by enumerating the necessary tools, giving the average price of a few of them, take up the boilers and their fittings, with engine details, show how to work up the separate parts from their castings, fit these together, and erect a horizontal engine and test it under steam. All this is minutely described, as it forms the groundwork for the construction of all the other engines. I next give examples of the different types of engines, Stationary, Locomotive, and Marine, as well as how to make a carriage and a model railway. After giving a few notes on pattern-making, proportions of engines, property of steam, etc., I conclude with a hot-air engine, which is easily fitted up. I have omitted oscillating cylinder engines, such as are sold in large numbers in the shops of the model-makers for boys to play with, as these are really models of no existing type of engine, merely toys; the making of them serves no educational purpose, and when made they are not at all satisfactory as regards their steaming powers. All the engines here have double-acting fixed slide-valve cylinders, and approximate in construction to those in actual practice. If carefully made according to instructions, none of them will fail to work, and it would be possible to demonstrate the action of the steam-engine to a class from any one of them, which really makes a model engine of some use after being made, as well as a source of pleasure to the maker.

The locomotive boiler described on page 37 will raise plenty of steam, as it maintains a good draught of air through the fire-box and fire-tube when at work. I have omitted the marine engine (as made at the present day) with circulating-pumps, air-pumps, and surface condenser, as being too complicated, but give two different types of compound non-condensing engines, which will afford some scope for the amateur's ability.

As regards the working drawings, these have all been carefully re-drawn by a competent draughtsman from original sketches drawn by me. Four of the sheets contain drawings of engines designed and constructed by myself, viz. the locomotive and tender, the bogie-tank locomotive, the centre crank horizontal, and the traction engines.

All these drawings may be thoroughly relied upon as being correct, and in one or two instances, by doubling the dimensions over all, a larger and more powerful working engine can be made.

Certain people maintain that model engine-making, like some other "hobbies," is simply a waste of time, and nothing more. Such persons, I am sure, are ignorant of what they are talking about, and I feel certain that most of my readers will agree with me "that there are few pastimes better fitted to give such good training to the mind, the eye, and the hand at one and the same time as model-making." For young Mechanical Engineers, during their apprenticeship, it is invaluable. All who can afford it should have a small workshop of their own, fitted with a set of tools, including a good lathe, so that they can devote their spare time to this pastime. For the same class who cannot afford to erect a workshop and procure tools, it might be advantageous if the Heads of large engineering establishments were to allow their apprentices and young engineers the use of the tools in the workshop (subject to certain regulations) during evenings and spare hours, for them to keep their hands in by making small engines, and perhaps supplement this by giving prizes for the best constructed and the best working engine under steam, made by their apprentices. By so doing, many lads would be kept out of mischief and temptation, be encouraged to follow their profession or trade, become dexterous in the use of different kinds of tools, get intimately acquainted with the mechanism of the several parts of the steam-engine, and acquire knowledge that would not be easily forgotten, and which would be of great service during after-life.

It has been suggested to me that if a Postal Amateur Mechanics' Society was started, on the same lines as the Postal Photographic Society, for the interchange of patterns, ideas, etc., between amateurs living at a distance from the large centres, such society would bring persons of kindred tastes into communication with each other.

J. A.

CONTENTS.

PART I.

BOILER AND ENGINE DETAILS, WITH TOOLS.

PART II.

DIFFERENT TYPES OF ENGINES: STATIONARY, LOCOMOTIVE, MARINE.

LIST OF ILLUSTRATIONS.

LIST OF WORKING DRAWING SHEETS.

The above are also published in a separate form.

PART I.

BOILER AND ENGINE DETAILS, WITH TOOLS.

MODEL ENGINE CONSTRUCTION.

CHAPTER I.

TOOLS REQUIRED IN MODEL ENGINE-MAKING.

THE **lathe** is an indispensable tool, and cannot be done without; but it requires some little practice to be able to use it properly. Let the amateur purchase a 3½″ centre bench lathe from any of the makers, but without back gear, as very often this is not used. The price varies, but we think that a good one may be had from £3 10*s.* up to £6, according to finish, and the number of chucks, etc. that go with it. Along with each lathe is usually supplied a drill- and driver-chuck, a hand-rest, and two tees. A face-plate and slide-rest are extras, and must be ordered if required, though we propose to do without them here. A $3\frac{1}{2}''$ centre lathe is a suitable size to purchase, as it is fairly strong, and will turn tolerably heavy work, and fine work quite as well, *i.e.* if the lathe is well made, the centres truly in line, and properly hardened, so as not to wear away rapidly. A lathe (Fig. 1) consists of two supports, firmly connected at their base, and fastened at right angles to the iron bed (which is 30″ long, and planed quite true along the top) by means of screws, A. The outer poppet-head is bored for the screw, and the inner one is fitted with a collar, within which the mandrel that carries the speed pulleys turns; the left end of the mandrel is concave, so as to allow of the steel point of the screw being tightened up and fit closely. C is called a T-rest; this slides along the slot in the bed, and can be clamped at any point, as well as elevated or depressed, when required. This is used for leaning the cutting tool upon, in order to afford greater steadiness when working. B is the right-hand poppet-head or tail-stock, which is also movable along the slot in the bed, and capable of being fastened anywhere, to suit the length of work

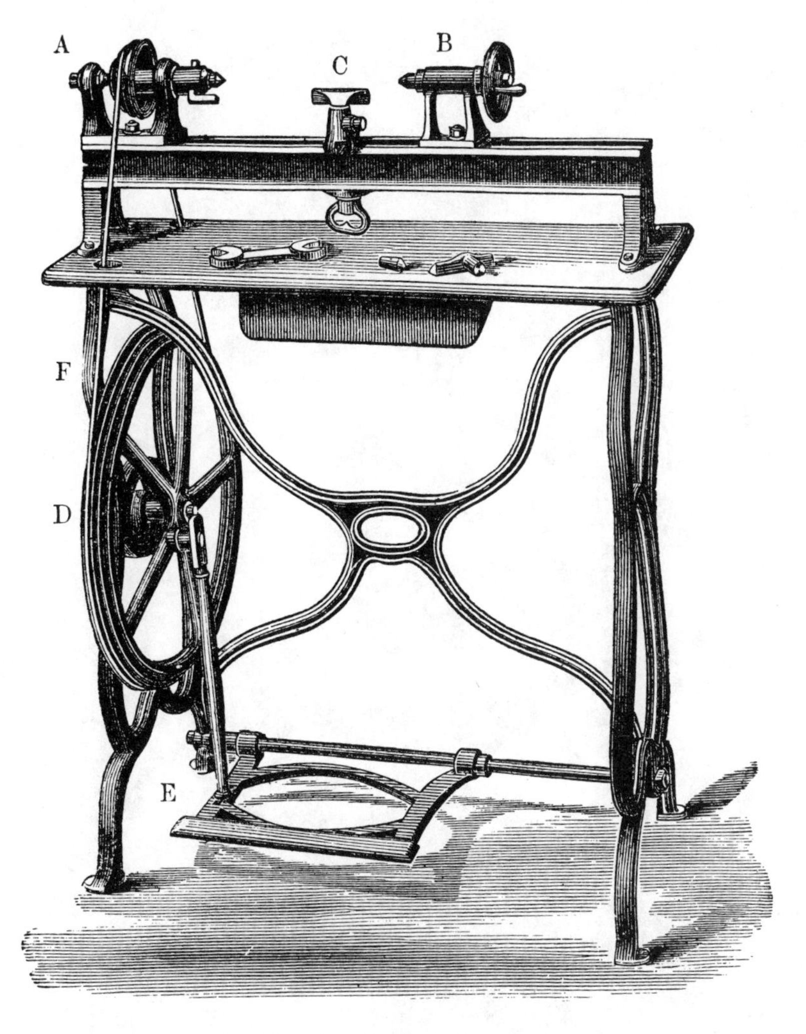

Fig. 1.—Bench Lathe.

A, outer poppet-head and speed pulleys; B, tail-stock or live-head; C, T-rest; D, crank-axle and fly-wheel; E, treadle; F, gut-band.

between the centres; its point can be advanced or drawn back, as required, by means of the screw and hand-wheel. D is the crank-axle which is connected with the treadle E by means of a rod, and so turns the fly-wheel, and this by means of a chain, or gut-band F, connects with the speed pulleys communicating motion to the mandrel. The pulleys on the spindle and mandrel are three or four in number, of different sizes, and so arranged that when the band is placed on the left-hand pulley a very rapid motion is given to the mandrel, as when turning wood. This motion is reduced more and more as the band is transferred to the right, till, at the extreme right, the rotatory motion is much slower than that of the spindle, and is used for iron-turning. There is generally a three-speed fly-wheel on the driving-axle as well. When the lathe is used for turning, and work must be centred in it, certain contrivances made of iron called "chucks" are fixed to the mandrel on which the speed pulleys are placed, and are for the purpose of centering the work, and making it revolve when the lathe is driven. Castings for making lathes are sold, but we advise the amateur to purchase his lathe ready made, for unless this tool is well made, with a perfectly flat bed, the points properly hardened, and standing in line facing each other, it will never turn out good work, even with the most experienced workman.

For metal-turning, a **driver-chuck** is used. This screws by one end on to the mandrel, and carries

FIG. 2.

a steel centre-point at the other; a steel rod which is generally bent at right angles outside, so as to be parallel to the line of centres, passes through the bore of the mandrel, and is fixed in position by a set-screw. Against this rod, the shank of the lathe-carrier (attached to the work) rests, and is taken round with it when the lathe revolves (see Fig. 2).

A **wood-chuck** is similar to the above, but carries

a three-pronged fork at one end. This is held in the chuck with a set-screw, the fork can be driven tightly into a piece of wood before placing it in the lathe; the prongs grip the wood, when the movable head is screwed up tight, and cause it to revolve with the lathe (see Fig. 3).

A **drill-chuck** is much the same as the last, having a hole at one end with a set-screw into which

FIG. 3.

different-sized drills can be inserted; some chucks have an extra plug for taking very fine drills. The wood-chuck can be made to act as a drill-chuck, by removing the fork, inserting the stem of a drill instead, and tightening it up with the set-screw (see Fig. 4).

Lathe-carriers are made of steel or iron; the head has a slot into which work must be firmly driven after centering, and tightened up by the set-screw. The stem, when the work is in the lathe, rests against the crank of the driver-chuck, and revolves, taking the work round with it. It is best to have two or three different sizes of lathe-carriers, and any blacksmith will make them (see Fig. 5).

A **bell-chuck** (see Fig. 6) is necessary for centering work to be drilled, as when boring out a cylinder. It is screwed on to the lathe, and the work, having been previously centre-punched, is put in the lathe, and retained in position by means of half-a-dozen

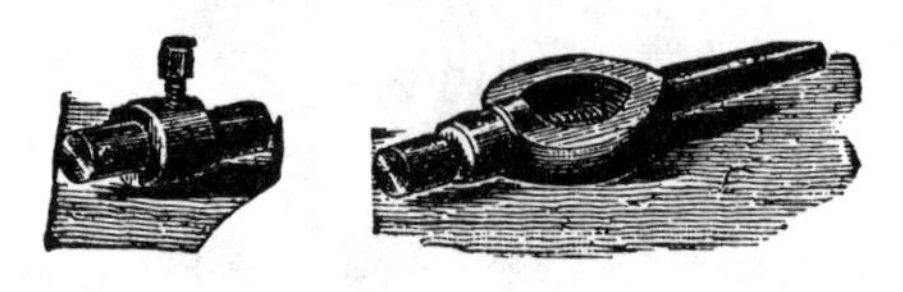

FIGS. 4, 5.

set-screws, which are tightened up equally all round the work, so as to keep its centres in line with those of the lathe. The advantage of this tool is that, having a bell shape, the drill can be sent completely through the work and out at the other end, which cannot be done so well with a face-plate. The accurate centering of work in a bell-chuck is very difficult, and takes some practice to do it properly. Price 14*s.* to 30*s.*

A **lathe-dog** is a contrivance which is useful for gripping fine work, and holding it in the lathe. The work is gripped between the two halves, the

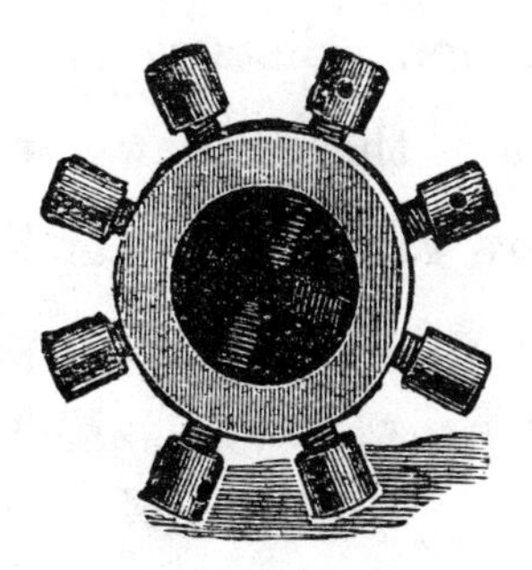

FIG. 6.

screws serving to draw them together; the tail can be bent, or left straight at one end to catch against the driver-chuck; if bent, it can be passed through a hole in the face-plate when chucked (see Fig. 7).

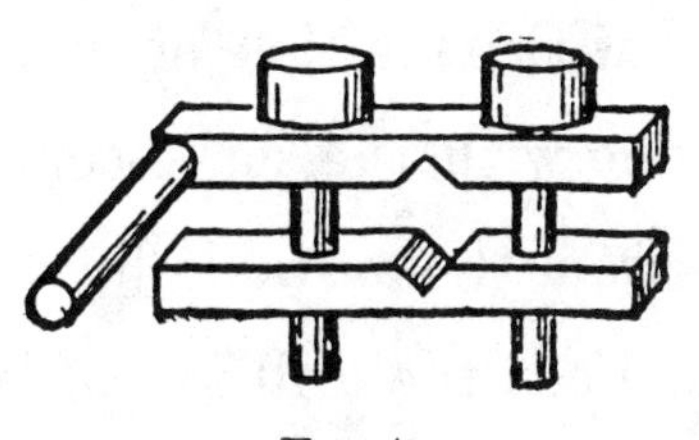

FIG. 7.

The **slide-rest** is a very expensive addition to a lathe (costing fully £3 for a $3\frac{1}{2}''$ centre lathe), and we will dispense with it for our work; with a little practice it can readily be done without.

A face-plate (Fig. 8) is a very useful adjunct to the lathe (costs 10*s.* to 15*s.*). It is an iron disc, planed quite flat on one side, and screws on to the mandrel by the other. Slots radiate outwards from the centre; through these screws or clamps "lathe dogs," as they are called, pass, and by them work is clamped to the face-plate, and properly centred in the lathe. The face-plate revolves and carries

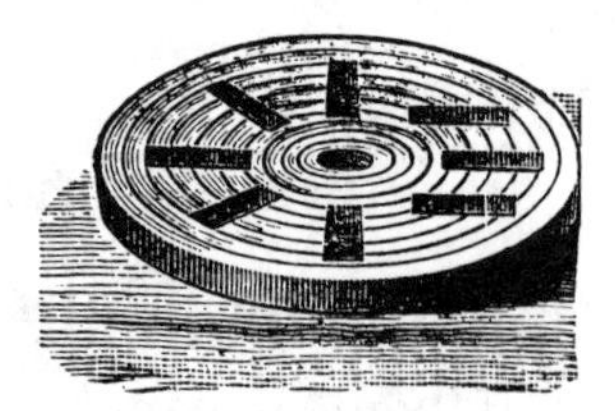

Fig. 8.

the work round with it, which should be so placed as to turn in line with the lathe-centres. Remove the steel point from the movable head, insert a suitable sized drill. Feed the drill forwards, by gradually pushing it up against the work by turning the hand-wheel, and it will soon cut its way through, if kept from revolving. This is done by clamping to the drill-stem, at right angles, a small iron bar, forked, or with a ring at one end, through which the drill passes; there is a set-screw for tightening up. The bar rests by its opposite end on the top of the T-rest, along which it slides with the drill. In this method, the work revolves and the drill is stationary, which gives the best results, but it is a little difficult to chuck the work properly at first. Work must always be first correctly centred, and then centre-punched, before attempting to chuck it in a lathe.

A **rose-cutter** is useful for finishing a cylinder. After boring out the cylinder, while it is still chucked, fix this tool to the mandrel and feed by the hand-wheel. When it has gone through the cylinder, this latter will be found to be quite circular inside, and as smooth as glass. Different-sized cutters are required for different-sized cylinders.

To drill a hole through a piece of brass.

Besides the above method, we have found the following plan give satisfaction. In this method, the drill revolves and the work is kept stationary, and is used in all cases where a hole is required to be drilled through the metal, but the keeping exactly between the true centres is not of the same vital importance as when boring out a cylinder. Screw the wood-chuck on to the poppet-head, remove the fork, and insert a drill, punch a centre hole at both ends of the work before chucking, remove the point from the movable head, keep the work from revolving, by gripping it with a pair of gas-fitters' pliers, or if larger, by clamping on a lathe-carrier, which will slide along the T-rest. Run the mandrel up against the work, and push the metal forwards on to the revolving drill by turning the hand-wheel. The drill will soon cut through the metal. The end of the work that rests against the mandrel must be filed flat before chucking, and be at right angles to the hole that is to be drilled.

Note.—A $3\frac{1}{2}''$ centre lathe will take the finest drills up to $\frac{5}{8}''$, for boring out brass or gun-metal.

A set of **wire-chucks**, capable of taking various-sized drills, from $\frac{1}{16}''$ up to $\frac{3}{8}''$, will be found very useful, especially if they are self-centering.

When turning metal, take care to chuck the work firmly; it can then be turned down accurately to the required form (using a pair of callipers properly adjusted, to give the correct size in diameter) with a graver or steel chisel, and finished with a fine file and oil, or emery cloth and oil, while still revolving in the lathe.

After the metal is chucked, it is first "roughed down" with a graver to the required form; second, smoothed with a chisel; third, polished with emery cloth and oil, and then removed from the lathe.

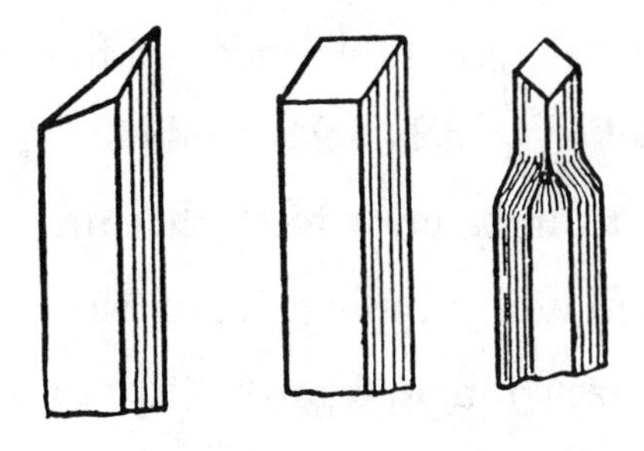

Fig. 9.

Be careful always when turning to use well-sharpened chisels; a grindstone is necessary to keep a good edge upon them.

Chisels are of various shapes: some have a flat edge, others are bevelled, and others again taper to a sharp point at one end. Of the last variety, it is best to have a right and left-hand tool. Chisels should be made of good square cast steel; old files do fairly well for this purpose—any blacksmith can make one out of an old file (see Fig. 9).

A screw-cutting lathe is not required for our work, as all the screws we need can be made by hand, with a set of taps and a screw-plate.

Vices. These are for holding metal when it is being filed down to lines previously marked upon it. We advise a 4″ jaw vice for holding heavy work, and a small 2½″ jaw table-vice for fine work, which can be attached to a bench or table. A small hand-vice, with a thumb-screw for tightening up the jaws, is very useful for gripping small work when filing. The small sizes range from 6*s.* up to 10*s.* 6*d.*, and the larger sizes from £1 10*s.* up to £5.

Files are sold in different shapes, and degrees of fineness in cut. A rasp is a very coarse file for roughing down wood; "rough" files are used for shaping,—"bastard" files may be purchased of any shape,—"smooth," "dead smooths," the finest of all, complete the various forms, along with the "triangular" and "round" files.

When using, fix the metal firmly in the vice, take the handle of the file in the right hand, press downwards with the left thumb on the tip of the file, pass it slowly and deliberately, but lightly, over the work, so as not to jump, and on the return stroke scarcely touch the work at all. The vice should be raised so high as to allow of the elbow passing over the vice when the arm is against the body, in order to use the file properly.

Surface-plate. This is a very expensive article, and is a block of iron the surface of which is planed perfectly flat, and is used for grinding the faces of cylinders, slide valves, etc., by the aid of emery powder and oil. It can be done without, and an iron for ironing clothes used instead. Get a perfectly flat smoothing-iron, cut off the handle, and fix it upside-down in a block of wood, and on the flat sole of the iron grind all the cylinders, valves, etc.; this does very well. Another way is to use a part of the flat bed of the lathe to grind the metal upon. The price of a surface-plate is from 8*s.* to 10*s.*, according to size.

Centre-punch. This is indispensable for metal-work, and consists of a piece of round steel tapering down to a point at one end, for marking out work, etc.; price 10*d.*

Broaches or **rimers** are useful for enlarging circular apertures, as in cylinder covers.

Screw-plate and taps. A watchmaker's plate and taps, from $\frac{1}{16}''$ to the finest sizes, are useful for delicate work. Another set of plate and taps, $\frac{3}{32}''$ to $\frac{3}{16}''$, is also required. These should be of the Whitworth Standard Gauge, and will be found quite sufficient for all our purposes. Prices run from 3*s.* 6*d.* up to 7*s.* 6*d.*

When it is desired to make a screw, take a piece of iron wire thicker than the size required, cut to the requisite length, leave a part longer than the full length of the screw for the head, grip the wire by one end in the hand-vice; having separated the jaws of the large vice a little way, let the wire rest between the jaws, leaning on their upper surface; take the hand-vice in the left hand and a file in the right hand, then gradually turn the hand-vice slowly round, keeping the wire resting on the upper surface of the vice; file away, gradually passing round the wire, till it gets of a smaller diameter than the part held in the hand-vice—this part forms the head of the bolt. Continue filing until the stem is rather a tight fit in the hole of the screw-plate that is selected for use; next grip the screw in the bench vice by the head; take the screw-plate, oil the hole, and gradually force it on by turning from right to left, when it will be found that a screw-thread forms on the wire. If the screw-plate when turning feels a tight fit, it is best to unscrew it and file down the stem a little bit thinner, or it is apt to break in the screw-plate and give trouble. After the screw is made the requisite length, remove the bolt from the bench vice, then you can file the head square, and shorten it also.

To make a nut, take a piece of brass, file it square, centre punch, fix in the vice, and with the hand-drill, drill a hole through it, suitable for the size of tap corresponding to the hole in the screw-plate that was used for making the thread on the bolt; grip the nut in the vice, and if a small tap, fix it in the hand-vice, oil it, and force it through the nut, while rotating the hand-vice in the right hand. This forms a thread in the nut of the same size as that on the bolt. When done you can file away the corners of the nut and make it octagonal, if you choose. If properly made, the nut will fit the screw on the little bolt. The above method applies to any size of bolt and nut it is desired to make; but for taps of $\frac{1}{8}''$ and larger sizes, instead of holding them in the hand-vice when using, tap wrenches with adjustable jaws, to suit different-sized taps (which give good leverage), are required to take them round; brass wire is not good to make small-sized screws from, as the thread when made is very apt to give way or "strip," as it is called. A little practice will enable an amateur to make all the screws and nuts he requires. This plan we recommend, rather than buying them ready made.

Pliers. It is best to have one or two pairs of different sizes, as well as a sharp cutting-edged pair for dividing wire, and a watchmaker's pair for delicate work. These cost about 1*s.* 6*d.* each.

Burnishers are highly-polished pieces of steel for polishing metal. They are rapidly moved over the part to be made bright; a little weak beer facilitates this operation.

A **frame-saw** is used for cutting metal, and is furnished with a thumb-screw to give varying degrees of rigidity, according to the nature of the work done. A good steel "tenon" saw does very well for cutting brass. Prices vary from 4*s.* to 5*s.* 6*d.* Saws for the above cost from 6*d.* to 10*d.*

Grindstone. A hand one is best for the amateur. Always use it with water, to prevent softening of the tool when sharpening chisels. These can be purchased. 15″ × 2″ is a suitable size.

Oil-stones are required for sharpening the tools after grinding their edges.

Iron block. A big block of iron, which can be firmly gripped in the bench vice, will be found useful for finishing off flat work, as bed-plates, locomotive frames, etc. Instead of this, an **anvil** would do.

Callipers. A pair of these is required for getting the correct dimensions of work when turning; those for outside measurement are the most useful. Price of a pair, about 1*s.*

Steel chisels, suitable for making steam-ports in small-sized engines, can be obtained at the model-maker's, or made from a steel bar, $1\frac{1}{2}''$ long and $\frac{1}{16}''$ broad at cutting edge; this size is useful for fine work, but it is well also to have a larger size, at least 2″ long and $\frac{3}{16}''$ wide at cutting edge.

A **pair of compasses** with a set-screw are required, price 1*s.* 8*d.*

A 36″ **foot-rule** is a convenient size to have.

Metal shears are required for cutting sheet brass and copper, price 2*s.*

A small **hammer** for riveting is useful, price 1*s.* 3*d.*

A metal **square** called an "engineer's" square is necessary, price 1*s.* 6*d.*

A set of **turning chisels** and **gouges** are required for wood-work and metal-work. Sets of these can be purchased in blocks, at prices from 10*s.* to 17*s.* 6*d.*

A **lacquering plate** is a plate of iron to be heated over the fire. Work to be warmed for lacquering is laid upon it.

Lacquering. Bright brass when finished must be lacquered to prevent its getting tarnished. Brass lacquer can be purchased, and is applied with a small brush; the metal having been previously heated in the spirit-lamp just hot enough for the hand to bear.

Brazing is rather too difficult for an amateur, and we will not describe the method here. For the engines given in this work no brazing is required, except for the internal fire-box and fire-tube; but it is best to have the fire-box made by a coppersmith, as will be shown further on.

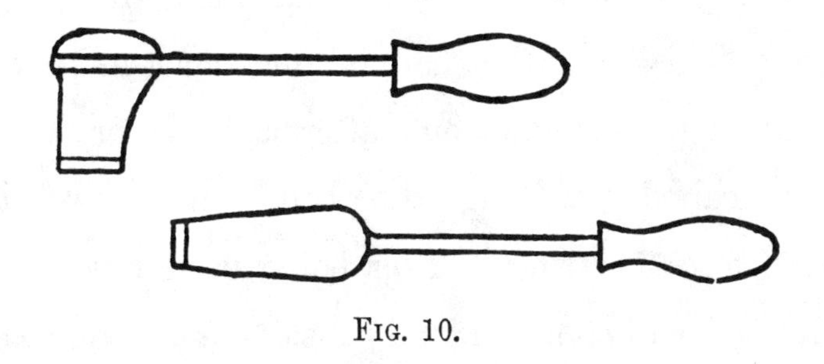

FIG. 10.

Soldering-bolts (see Fig. 10). Soldering is required for model-making. This can be easily done and joints made perfectly steam-tight, after a little practice. We recommend two light copper bolts: a straight one, and one bent at right angles to the handle; and for a flux do not use resin but spirits of salt, made by saturating hydrochloric acid with strips of metallic zinc, till no more hydrogen is evolved; this forms zinc chloride. Keep the solution in a well-corked bottle. By means of a wire touch the parts over with the solution before soldering, after having scraped them clean.

Method of soldering. Take the soldering-iron, warm it in the fire, with a file clean it thoroughly near the point on both sides, and if at the proper heat, after rubbing a little spirits of salt upon it, and touching it with a stick of soft solder, the solder will stick to the iron, and make it appear bright. This operation can be done at the proper heat (which is found out after a little practice), and is called "tinning the bolt," and must be done before soldering. After being used much, the "tinning" will come off, and must be again renewed by the same process. After "tinning," warm the soldering-bolt in the fire; never let it arrive at a bright red heat, as this is too hot for use. Before beginning to solder, lay a stick of solder at one side, and a bottle with spirits of salt at the other. If you are going to solder two pieces of brass together, first clean both pieces by filing all round, where they are to join; sprinkle a little spirits of salt on as a flux, to remove the tarnish; then, with the soldering-bolt touch the stick of solder lightly, when a piece will adhere to it. Put the bolt with the solder upon the brass over the joint, and hold it on for a second or two, when the solder, if all tarnish be removed, will adhere to each piece, and they will become tinned over like the bolt. After tinning, bring the two pieces together, put on more spirits of salt, touch the solder-stick again with the bolt, and hold it over the pieces of brass for a little time to "sweat" in the solder; when cool, the two pieces will be found to adhere firmly together. At first, one is apt to make a clumsy job, and put on too much solder, but a little practice will enable the amateur to make a neat job and a first-rate joint. Brass, as well as iron and copper, must be tinned before attempting to solder two pieces together. Tin plate does not require tinning. Iron is more difficult to solder than brass, and for copper, the soldering-bolt must be made a little hotter than for brass.

Metals. Brass wire is generally hard, and can be softened by being made red-hot. Castings are the softest, and can be rendered hard by hammering, and soft by being made red-hot. It is best always to use new files for this metal.

Gun-metal is much harder than brass to work.

Copper is softer than brass.

Iron is much harder than brass. Water as a lubricator assists in turning and sawing; in drillings it requires oil. It takes the cut off files rapidly, and old files will cut it just as well as new ones.

Note.—Never use the same files for brass and iron, but keep separate tools for each kind of metal.

Bright brass may be finished with emery powder and oil, the burnisher, or by dipping. This last consists in making the work free from grease, and then "dipping" it clean and warm into nitric acid (aquafortis), and when it has "taken," rinsing rapidly in clean cold water and drying quickly; this makes the work beautifully bright.

To temper drills. The small-sized drills had best be purchased, or made out of steel knitting-needles; the larger sizes can be made out of a steel bar by a blacksmith; these must be each ground to a point and then tempered, when they are ready for use.

Heat the drills (use a Bunsen burner for the small sizes) in the fire until red-hot, about the colour of red lead; if small plunge them into oil or cold fat; if large, into water; clean them bright with emery cloth, again heat them until the edge polished is a golden or straw colour, then quench in water, and when ground they are ready for use. If the steel be brittle, let them run lower to a blue colour. A blow-pipe is good to get the proper temperature with when acting on a gas-jet.

Fine brass tubing is sold in foot lengths, each length costing 6*d.* The finest is $\frac{1}{16}''$ bore, the next $\frac{3}{32}''$ bore, and the next again $\frac{1}{8}''$. For larger sizes ordinary gas-tubing does well enough.

A **hand-drill** is necessary, one that takes drills from $\frac{1}{4}''$ to $\frac{1}{16}''$ is a useful size; the chuck is adjustable for the different sizes of drill-points, and these are supplied along with the tool. When using it, the drill is kept pressed against the work with the left hand, and the wheel that causes the drill to revolve driven by the right hand. Prices, 5*s.* 6*d.* up to 12*s.*

An upright **vertical drill** for fastening to a bench

is useful for all sizes of drills up to $\frac{1}{2}''$. Prices range from £2 up to £2 10*s.*

Drill-points can be obtained from Mr. Lee, 203, Shaftesbury Avenue, London, who supplies one dozen, varying in size from $\frac{1}{16}''$ up to $\frac{3}{16}''$, for the sum of 1*s.* These drill-points are very good ones.

A good illustrated catalogue of "American" tools, with prices, can be also got from Mr. Lee, and should be in the hands of every amateur.

In case the amateur should prefer to make his own drills, we will conclude this description of the tools by saying that very good small-sized drills can be made from steel knitting, and sewing-machine needles. The former can be cut into lengths of $1\frac{1}{4}''$, and a drill made out of each length; one drill can be made from the latter (broken needles do for this purpose). File the steel wire slightly tapering at one end, and then spread the point with a good strong blow from a fairly heavy hammer (a series of light blows will not do at all), grind the point to a sharp cutting edge on the grindstone, and the drill is ready for use—or if it requires to be tempered, instructions have already been given how to do this operation.

A word or two about a **workshop** may not be out of place. A spare room in the house may be utilized as such, or a wooden structure may be erected outside. Whichever is adopted, the workshop should be well ventilated, free from damp, have plenty of light admitted, and be tolerably warm. A fire-place or small stove is useful for heating soldering-bolts in, and warming the place during winter. The shop must be well stocked with engineering tools, and these should be arranged in some kind of order, so as to have "a place for everything." The lathe ought to be fixed near the window, so that a good light is thrown upon the work when turning; all the turning tools may be conveniently kept in a box fixed to the lathe-stand. A joiner's bench is necessary, as well as a few tools for wood-working and pattern-making. A forge can be dispensed with, but if there be a supply of gas, a Bunsen burner is useful, for tempering fine-sized drills, etc. A vice, with 4″ or 5″ jaws, should be fixed at a convenient height from the floor, to suit the operator, as well as a smaller size for holding fine work when filing. A grindstone ought to be fitted up, so that it may be driven by hand, or preferably by foot, and is useful for grinding tools, as wood-cutting and turning tools must always be kept very sharp; and there should be some arrangement to keep the stone moistened with water when in use. Perhaps the best way is to keep it half immersed in a water-trough, and the water can be run out of the trough by a cock when not in use. If the amateur goes in for locomotive building, it is a good plan to fit up a permanent length of railway, fixing it on a board near the wall, on one side of the shop. This can be raised to any convenient height, and is always ready for running the locomotives upon, when it is desired to do so.

For **pattern-making tools** see page 308.

A few of the electros for reproducing some of the drawings of the tools have been kindly lent by the Britannia Co., Colchester.

CHAPTER II.

THE BOILERS AND THEIR FITTINGS.

WE first describe a horizontal boiler without a flue-tube, suitable for hanging over a kitchen fire, and useful for testing small engines when making them. Next, a locomotive boiler with an internal fire-box, fire-tube, and a water space all round the fire. This boiler raises plenty of steam, as from its construction a strong draught of air is kept rushing through the fire-box, fire-tube, and funnel when at work. Lastly, a vertical boiler, constructed in a similar way to the locomotive boiler.

The **horizontal boiler** (see Fig. 11A) is made entirely of No. 20 sheet brass, riveted and soldered together, and when complete is 10″ long and 6″ in diameter. Cut a sheet 19″ × 10″, and bend it into a circular shape, with a diameter of 6″; a sheet this size allows of an over-lap of fully $\frac{1}{2}$″. Clean the ends that overlap with emery paper or dilute nitric acid; this makes them take on solder when chloride

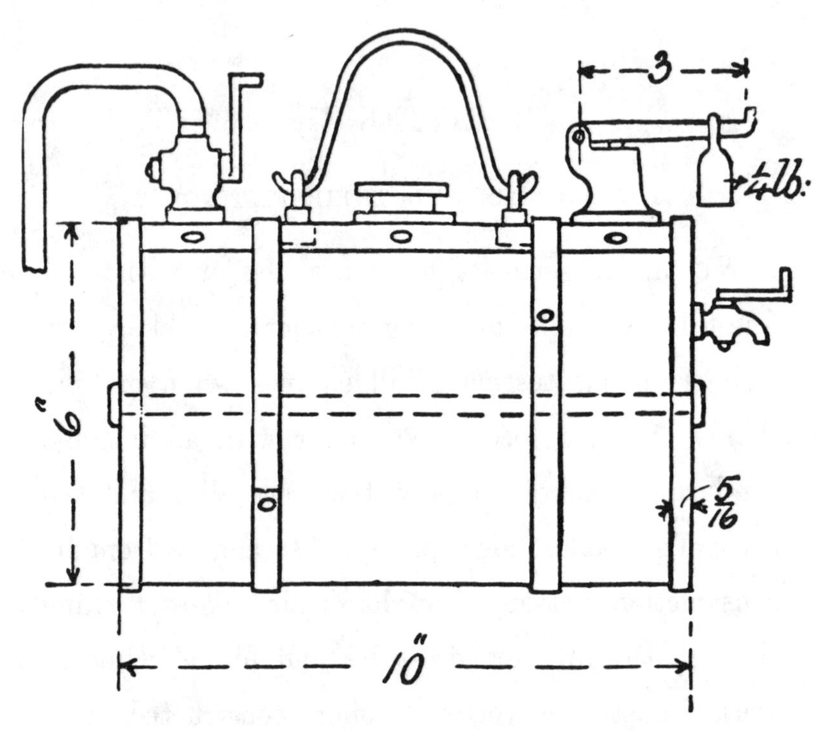

FIG. 11A.—Boiler for Kitchen Fire ($\frac{1}{4}$ full size).

of zinc is applied. Fasten these ends together along the longitudinal seam here and there with solder, then drill three holes for rivets which pass through the over-lap, one near each end, and one near the centre; put through three small copper rivets from the inside, and rivet them firmly on the outside. The end rivets may be fixed against a leg of the vice when riveting; but for the centre one it may be necessary to pass the boiler over an iron block held horizontally in the vice, and then rivet while the head of the rivet rests on this block. Solder the shell firmly together along the seam and over the rivet heads. For the *ends* cut out two circular pieces of sheet brass, each $6\frac{5}{8}''$ diameter; this size allows of a $\frac{5}{16}''$ collar all round, which is to be turned down over the barrel. The collar is made by hammering the end over a circular block of hard wood, turned to the same diameter as the barrel, and long enough to be gripped in the vice; tin the inside of the collar with solder, and also a circular strip around the edge of the barrel at both ends, place the end in position, and solder it firmly to the barrel; do the same with the other end. Pass a stay through the boiler, and for this purpose drill a small hole through the centre of each of the boiler ends; take a piece of brass wire $\frac{1}{4}''$ thick—brass is used, as this does not rust—and about 11″ long, tap a thread at each end, fully $\frac{1}{2}''$ in length, screw a nut on one end, cut the wire short, and hammer the edges all round down on the nut to keep it firm; solder it on, and pass this wire through the boiler from one end, and screw the other nut on to the opposite end; bring it up to rest against the end of the boiler, and solder both the nuts over. This stay makes the boiler strong enough to resist pressure. Arrange the boiler so that the longitudinal seam is placed near the top or upper part when the fittings are on. These consist of a $\frac{1}{8}''$ gas stop-cock, and a foot or so of tubing, a screw plug or water-filler that screws into a ring soldered on the boiler, and a lever safety-valve (with a weight), $\frac{3}{16}''$ diameter at narrowest part of seat; holes must be drilled in a straight line along the top of the boiler, and each enlarged with a rimer to the size suitable for the corresponding fitting; then solder on these fittings. A weight of about a quarter of a pound, hung at a distance of 3″ from the fulcrum, will be suitable for this size of valve. Cut two strips of sheet brass or copper, each 19″ × 1″, rivet them together to form two rings, of a size that will just slip over the boiler, one from either end, before the ends are put on, and be a tight fit; these can be fixed with solder in certain parts, and are used to give extra strength. Any blacksmith will make and rivet together a small iron frame, which will sit on the hob over the kitchen fire, or on the grate, and support the boiler when in use, but it must be so made that the bottom part of the boiler is exposed direct to the flames. Or if there be an arrangement in the fire-place of a swing with hooks, and it is desired to hang the boiler like a kettle, this is easily done by purchasing two strong rings such as pictures are hung by, file off the screw-thread from the stems, as it is too rough for our purpose, tap a new thread, and make nuts to suit, and all you have got to do is to drill two holes in the upper part of the boiler, one near each end, in line with the fittings, insert the stems of the rings, and screw the nuts on inside. Of course all this must be done before the boiler ends are fixed on, and any of the fittings attached. Take a strong piece of copper wire, bend it into a U-shape, insert the free ends through the rings, and twist them with pliers, so that they will not come out. By means of a hook passing through the wire loop, the boiler can be hung any distance above the fire. Solder a small cock into one end, to show when the boiler is about three-quarters full of water. When in use, temporary attachment is made between the boiler and any engine by india-rubber tubing (different sizes can be purchased), one end being pushed over the steam-pipe—which should project outwards for a considerable distance from the boiler, as it gets very hot for the operator if too near the fire—and the other end attached to the steam-pipe of the engine, which rests on a chair or table. To prevent the rubber tubing being blown from off the ends of the steam-pipes, it should be tied on at both ends. The disadvantage of this india-rubber method of attachment is, that if the pressure rises too high in the boiler, the rubber tubing will burst with a loud report and be rendered useless; so that it is perhaps preferable to make the attachments by means of coupling screws and brass tubing temporarily soldered together; different sizes of tubing can be joined together, so that though the boiler steam-pipe is $\frac{3}{16}''$ bore, that supplying the engine can be reduced, if necessary, to $\frac{1}{16}''$ bore when testing a small-sized engine.

Locomotive boiler (see Sheet No. 1). All the locomotive boilers we take up here are made on the same principle, though they vary slightly in size. The **boiler barrel** and **outer fire-box** is made of No. 24 sheet copper, the piece forming the barrel and outer fire-box is $10\frac{1}{8}'' \times 8\frac{3}{8}''$. There is an **internal fire-box** and **fire-tube** made of No. 21 sheet copper (the thinnest size that can be brazed). This article must be brazed together in all the joints, and is best made by a coppersmith, as brazing is difficult for the amateur. The fire-box is composed of three separate pieces, one bit forming the two sides and top; these must be flanged outwards at the foot. The back or fire-door plate has an oval hole cut in it, and is flanged at the foot. The front plate is also flanged, and has a circular hole cut in it for the fire-tube to pass through, there being *one* fire-tube only that goes to the smoke-box. The front and back plates must be cut rather larger than the dimensions given, to allow of a small collar being turned down over and brazed to the sides when putting together. A small oval sheet-copper tube is brazed together, and also brazed into the fire-door hole; this forms the coal-shoot or fire-hole ring, and must project outwards, so as to pass through the outer fire-hole when the boiler is put together. The fire-tube is made of copper brazed into the fire-box, and also into the boiler tube-plate, and the ends can be flanged down on the plates as well; the longitudinal seam is also brazed. When this fire-box is finished the barrel and outer fire-box must be bent into shape by hand, and fitted over the inner fire-box, the barrel being made of the correct diameter and soldered along the longitudinal seam, which is underneath, no rivets being required along this seam; flange the tube-plate over the front of the barrel all round, and solder them together, having previously tinned the parts. Cut out the outer fire-box back plate, cut a fire-hole in it, and pass the projecting tube from the inner fire-box through it; beat down the ends to form a collar—this must be done inside as well. The collar fixes the plate; solder it all round, do the same to the barrel after flanging. It will be noticed that the inner fire-box has a projecting collar and flange all round underneath, which ought to be very *carefully brazed* at the corners; for unless this is done, it is apt to leak at these parts. The use of this projecting part is, that when the boiler is put together, this collar projects outwards all round against the sides, the front, and the back of the outer fire-box, and forms the foundation-ring. The space between the fire-boxes is called the "water space." Solder the parts where they come together all round carefully, after tinning them to take on solder; sweat the solder well in between the joints to render the boiler thoroughly water-tight, and to do this the boiler must be turned upside down and fixed against a support of some kind—two blocks of wood do very well. Fit on the front plate of the fire-box by cutting out of it a piece of a circular shape, to form a concavity to fit the circle of the barrel, flange the sides, and solder together. The boiler is now ready for staying. We proceed in the same way as before. Three small copper rivets may be quite sufficient; two might do, one on either side, but it is best to put one in front between the fire-boxes as well. The back plate does not absolutely require a stay, as the fire-hole ring acts as one; still, it is best to put one through under the fire-hole (see Sheet No. 1). The fire-boxes must be firmly stayed together to stand pressure, as this is the weakest part in the whole boiler.

Be careful to insert the rivets below the level of the fire-door, as it is as well to have no openings above that level, for fear of the water getting short and the solder melting. The outer fire-box projects further downwards in front and at the sides than at the back, where it is cut off short; this is done to leave a space for the lamp to pass to the fire-box.

The **spirit-lamp** is made of tin, and the neck is left of such a width and length, that when it is in position in the fire-box a clear space of about half-an-inch is left between it and the fire-box, both in front and at the sides, for air to pass to the wicks and support combustion.

The **smoke-box** is made of tin of a circular shape, surrounding the barrel; a rivet is put through underneath where the ends overlap, to keep it circular. A front plate of tin, having a 1″ hole cut through it, can be flanged and riveted by very fine rivets made of copper wire to the smoke-box, as solder will melt by the heat. A small circular door, flat or dished,—the latter is rather difficult to do,—is hinged in front by a hinge made of sheet brass and a wire, which must be riveted to the smoke-box and to the door. This door opens and shuts, and when shut it fastens by means of a small brass handle. A thin strip of brass is riveted across the opening of the smoke-box inside, and a T-headed bolt passes through a slot in the brass; the flat of the head corresponds with the breadth of the slot. It is turned a quarter round by means of a handle, and then the outer handle is tightened upon the inner one in lock-nut fashion. The funnel is about $\frac{5}{8}''$ wide at the root, made of tin, soldered and riveted together. It is fixed on the smoke-box by a collar, which is fastened by two rivets, a hole being first cut in the smoke-box of the same size as the funnel over which it is placed; a bit of half-round wire soldered round the top will form the beading, but this is unnecessary. When placed on the barrel, the smoke-box overlaps the front of the boiler by about $\frac{1}{8}''$, and to prevent it from passing farther backwards, a copper wire is neatly soldered all round the barrel for the edge of the smoke-box to rest against, when it is pushed on to the front of the boiler over the tube-plate. The smoke-box should never be attached by solder to the boiler, because it is necessary sometimes to take them apart; but white lead can be put round the joint to keep it air-tight.

The **fire-door** is made of copper, oval in shape, and is hinged, the hinge being riveted to the fire-door, and soldered to the boiler. A small latch handle is pivoted to a rivet, and passes into a keeper when the door is shut; the keeper, of sheet brass, is soldered to the fire-box. This boiler, if properly made, will raise steam of 20 to 30 lbs. pressure per square inch, and stand it quite safely.

The **fittings** are a dome, with or without a lever safety-valve, a screw-plug for filling the boiler with water, a $\frac{1}{8}''$ spring safety-valve which will relieve the boiler of excessive pressure, and a small gauge-cock to test the height of the water in the boiler when filling it. The steam-pipe, $\frac{1}{8}''$ bore, passes down through the smoke-box to the cylinders, starting inside the upper part of the dome; the pipe is soldered where it passes out through the front tube-plate. Steam is controlled by a $\frac{1}{8}''$ stop-cock placed in the smoke-box, and into which the steam-pipe is screwed, and below the cock there is a coupling screw or union for further attachment by tubing to the cylinders. This cock is opened and shut, by simply drawing out and pushing in a hinged wire lever, or regulator (pivoted to the back of the fire-box), and this is done by means of a wire passing through a $\frac{1}{16}''$ brass tube, which extends through and is soldered to the boiler at the tube-plate and back of fire-box. This arrangement of the wire passing through a tube, when inside the boiler saves making two stuffing-boxes for the regulator-rod to pass through, as these are difficult to make and keep steam-tight when very small, and the tube serves the same purpose. A full-sized drawing of this arrangement is given on Sheet No. 11, B.

The spirit-lamp is the same as that on Sheet No. 12, on which will be found a plan and elevation. It is made of tin plate soldered together, the rectangular tank sits behind the fire-box; a long circular funnel of tubing or tin, with an elbow-joint below, is soldered at the back of the tank on one side, for the purpose of pouring in spirits of wine to the wicks when steam is up. The neck of the lamp is of tin, similar to the tank, but narrower, so as to allow a clear space of half-an-inch all round and in front, between it and the sides of the inner fire-box, for admission of air. The lamp can be pushed in and drawn out of the fire-box as required. It carries on the neck a number of wick-carriers, set in two rows; these are simply $\frac{3}{16}''$ holes drilled in the top plate and stuffed with cotton wicks, or asbestos twine. As a rule, we allow twelve or fourteen of these small wicks to burn in a boiler having a fire-box (measured below the bottom of internal fire-box) about $3\frac{7}{16}'' \times 2\frac{3}{8}''$, as fewer than these will not keep up sufficient steam.

Remarks.—The boiler will raise steam up to 30 lbs. pressure per square inch. Its method of construction was only arrived at after one or two experiments carried out on small boilers by ourselves, and we believe that its steaming qualities depend entirely on the following modifications.

1. Increasing the depth of the inner fire-box, by carrying the crown upwards to near the top of the outer fire-box, and having a circular crown, as this increases the heating surface.

2. Instead of making the boiler multitubular, adopting one fire-tube only, for in small boilers, if there be two or three tubes, the draught gets choked, and steam rapidly falls.

3. Discarding an ash-pan, and leaving the fire-box open underneath, with a space of nearly $\frac{1}{2}''$ between the sides, and those of the lamp, for admission of air.

4. Cutting the blast-pipes (there are two) short, so that their orifices open and discharge steam about the centre of the front of the fire-tube; their orifices are slightly flattened to increase the pressure. More steam is generated than when pipes are led up to the root of the funnel.

5. Adopting a wide funnel (one less than about $\frac{5}{8}''$ diameter is rather narrow, to make the boiler draw well).

6. Not firing the boiler with one or two wicks, as is done in some model boilers, but using a great number, so as to generate more steam than the cylinders can use.

The above boiler will drive a pair of cylinders each $\frac{1}{2}''$ bore and $1''$ stroke, or one cylinder $\frac{3}{4}''$ bore and $1\frac{1}{2}''$ stroke.

Note.—With such a number of wicks burning inside the fire-box, after a time the spirit-tank is apt to get a little hot and cause evaporation of the spirits (seen by fumes coming out of the funnel leading to the lamp), in which case fresh methylated spirits must be poured in to keep the wicks burning; but there is not the slightest danger of an explosion of the spirit taking place.

A much larger and stronger boiler can be made by doubling the dimensions over all of those given on Sheet No. 1, and using stronger material, while still adhering to the same method of construction. It is a mistake to make a boiler multitubular of a size less than a half-horse power, consequently we adopt one fire-tube as before. Make the barrel and outer fire-box from a single sheet of No. 16

sheet brass; the internal fire-box, fire-tube, and tube-plate from the corresponding copper sheet No. 18. When put together the boiler must be brazed at every joint, both at the outside joints and at those connected with the inner fire-box, so that it is best made by a coppersmith, to insure its being strong and able to stand 60 lbs. pressure per square inch. Pass eight $\frac{3}{16}''$ copper rivets through between the fire-boxes on each side, and in front and behind; after riveting together, their heads should be brazed over, both on the outside and inside of the fire-box. A longitudinal stay is not required. The fittings should be brazed to the shell, and these should include, besides a half-inch safety-valve, two gauge-cocks, a glass water-gauge, and a clack-valve for a pump. The smoke-box and funnel are made of sheet brass. A lamp can be made of thick tin plate, with wicks to burn methylated spirits of wine or oil; but a better plan is, to fit a brass grate low down inside the fire-box (leaving it all open underneath), which will burn small blocks of wood steeped in methylated spirits of wine, paraffin, or ordinary oil, and this will raise abundance of steam.

Vertical boiler (see Sheet No. 2). This form is easier to construct than the last. It is made of No. 24 sheet copper cut to a size of $6\frac{1}{2}'' \times 10\frac{3}{16}''$; this is bent into a circular shape, and forms the outer shell. The internal fire-box is made of No. 21 sheet copper, and should be done by a coppersmith, having all the joints brazed together. It is made in the same way as the previous one described, but is circular; there is a flat top or crown, through the centre of which the funnel passes. The fire-hole is cut to dimensions of an oval shape, and a piece of copper brazed in to form the fire-hole ring is left long enough to pass through the outer fire-hole when put together. When the fire-box is completed, take the sheet which forms the outer shell and measure off the distance for the fire-hole; make the outline oval in shape, drill out the piece, and file to outline. Now bend the metal into a circular shape to the correct size, then widen it a bit, and insert the fire-box from the lower end, the funnel passing upwards through the boiler. Flange the fire-door ring and solder it on the inside of the outer fire-box, bend the shell again to the correct shape, solder all along the longitudinal seam—a small rivet may be put through the overlapping edges near the top, and riveted against a small block of iron fixed in the vice, and which passes inside the boiler. Another small rivet may be inserted in line with this one, and must pass through the lower part of the inner fire-box below the water-line, where the metal is curved outwards to meet the outer fire-box, forming the foundation-ring, and connects together the two fire-boxes; the heads of the rivets must be soldered over, and solder be put in all round the joints of the outer and inner fire-boxes, and for this purpose the boiler should be turned upside down. An opening must be made at the foot under the fire-door $2\frac{1}{2}'' \times \frac{9}{16}''$, to admit air to lamp, and clean out the ashes. At one corner this opening may be enlarged to $\frac{3}{4}''$ (not shown in drawings), to allow the funnel of the lamp to pass through when *in situ*. This opening can be cut after putting the boiler together, by slitting it at the ends to the required depth with metal shears, breaking out the piece with pliers, and then filing to outline. The crown is made of sheet copper, $3\frac{7}{8}''$ diameter, to allow of a flange, which in this case is put outside the shell (instead of inside, as in large boilers), for the purpose of strengthening the boiler more when soldered. The funnel passes through the centre, and is soldered to the crown, or if preferred, the boiler crown could be brazed to the funnel by the coppersmith, and so increase the strength still more. A sheet-brass door —one plate only—is cut rather larger, but of the same shape as the fire-hole, attached by rivets to a hinge, which in this case must be soldered to the boiler; there is also a keeper and latch to fasten the door to.

We insert the following fittings: a $\frac{1}{16}''$ cock, so placed as to indicate when the boiler is three-quarters full of water; a second cock is put in at a height just under the fire-box crown; when working, while water continues to be blown out of the lower cock, the boiler may be safely run without fear of the water getting too short. The boiler will steam about twenty minutes with one charge of water. A small water-gauge can be made or purchased, and inserted, so that the water stands well up in it, when the boiler is three-quarters full. A $\frac{3}{8}''$ water-plug or filler is soldered into the crown, as well as a spring or lever safety-valve, fully $\frac{1}{8}''$ diameter at its seat. A cock controls the supply of steam to the cylinder of the engine, which can be attached by tubing and coupling-screws. This boiler will drive a $\frac{7}{16}''$ bore cylinder and $1''$ stroke very well. The boiler could be fitted with the force-pump (see p. 72), but this would require to be driven by hand.

The spirit-lamp is made of tin (see drawings). It consists of a circular tank soldered together, about $\frac{3}{4}''$ deep, and of a size that will fit easily inside the fire-box. On the top a number of small holes are drilled, which act as wick-carriers; about a dozen will be required to keep up steam in a boiler of this size, if each hole is $\frac{5}{16}''$ in diameter. Plenty of wicks must burn in these tubular boilers; to try to keep up an active supply of steam with one or two wicks only burning, as is done in most models, will render these boilers pretty nearly failures, for no strength of steam can be kept up, with an engine running at the same time. Spirits of wine is poured into the lamp through the vertical lamp-funnel when required, to keep the flames brisk when steam is up, for with such a big fire burning it requires constant stoking. The boiler is not fixed to a wooden stand, but is simply lifted off and set down upon the lamp resting on the table. When the fire

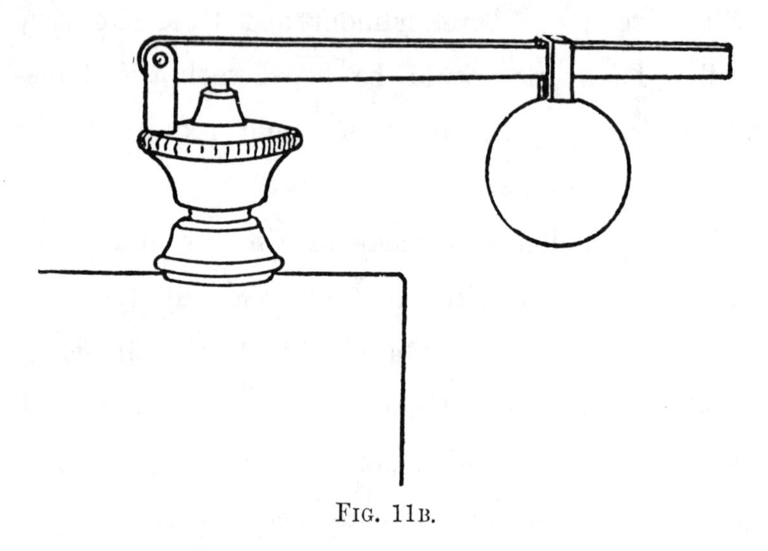

FIG. 11B.

is to be "drawn," it is best with a cloth to lift the boiler off the lamp, and then smother the flames with the same cloth rapidly brought down, so as to extinguish them. The flames do not blow out very easily, and not at all when in the fire-box, while to allow the methylated spirits to burn out is apt to char the wicks, and they do not stand long. A flange can be soldered or riveted all round the boiler at the foot to steady it when at work. The inner fire-box, being of a circular shape and brazed together, does not require to be attached by stays to the outer shell, as the circular shape tends to resist the pressure more than the flat-sided fire-box in the locomotive boiler does.

Boiler Fittings.

The **safety-valve** (see Fig. 11B) is a circular, steam-tight valve set on the boiler. Its object is to prevent the steam pressure in the boiler exceeding the working pressure. The pressure on the outside of the valve is produced by a lever and weight, or by a spring, as in the case of locomotive safety-valves. The valve casting is a brass casting turned on the lathe, and a central hole is drilled through it. At the top this hole is widened, and made of a conical shape, for a short distance downwards, and this forms the valve-seat proper. The valve itself is a piece of circular brass rod turned on the lathe, and made conical round the edge, so as to fit the seat in the casing; a small spindle, or stalk, is turned on the valve below the conical edge, which guides the valve in rising and falling into its seat. The valve must be ground with emery powder and oil into its seat, so as to be rendered steam-tight. A brass fulcrum must be slotted at one end, and screwed by its other end into the rim, or flange of the valve-casing, and a piece of steel or iron wire, filed or hammered flat to form a lever. This is drilled at one end to take a pin, and by it is pivoted in the slot of the fulcrum. In some cases, this lever simply rests on the top of the valve; in other cases, the top of the valve is prolonged upwards above the casing, and the lever is pivoted to it, as it passes through a slot which has been made in the top part of the valve. The weight, which is hung over the free end of the lever, is a piece of brass, iron, or lead; the projecting part by which it hangs must have a slot cut in it, so as to be able to slide the weight along the lever. A hole can be drilled through the slot, and a set-screw attached, which, when tightened up, will keep the weight fixed at any part of the lever, as desired. The valve-casing should be soldered or screwed into the boiler.

A **spring safety-valve** (see Fig. 12) is similar to the above, but, if of the lever type, the end of the

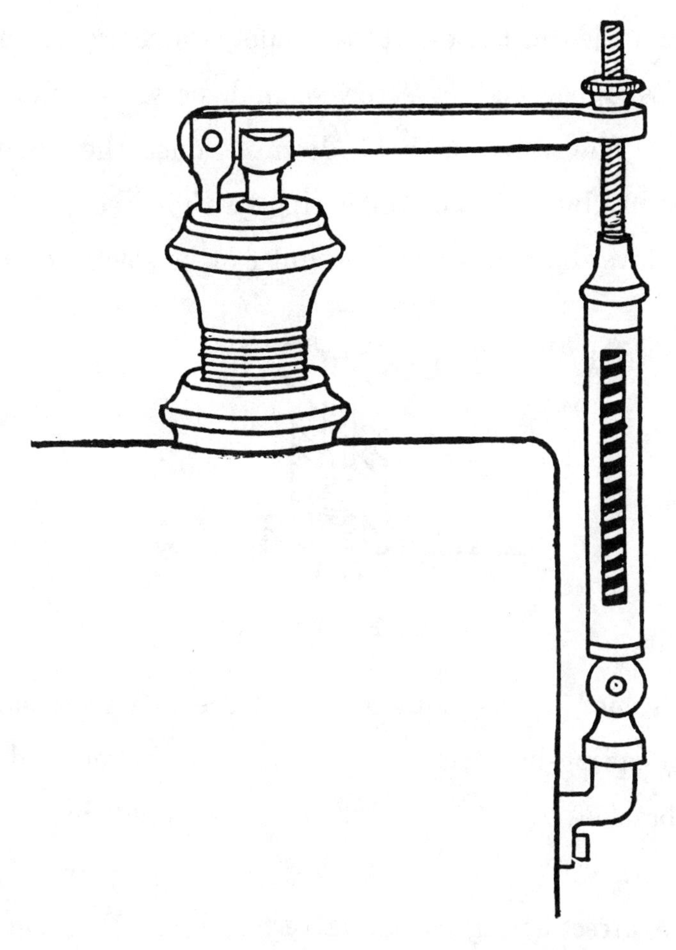

Fig. 12.

lever is pulled down by a Salter's spring-balance, which is generally graduated. The balance can be made with a small spring acting against a piston, moving up and down in a small tube, closed at both ends. The piston-rod passes vertically through a hole drilled in the end of the safety-valve lever, and at its upper end is screwed, and provided with a nut. The valve is held down against the steam pressure by means of the spring in the tube, pulling the lever downwards. The nut can be slackened, or

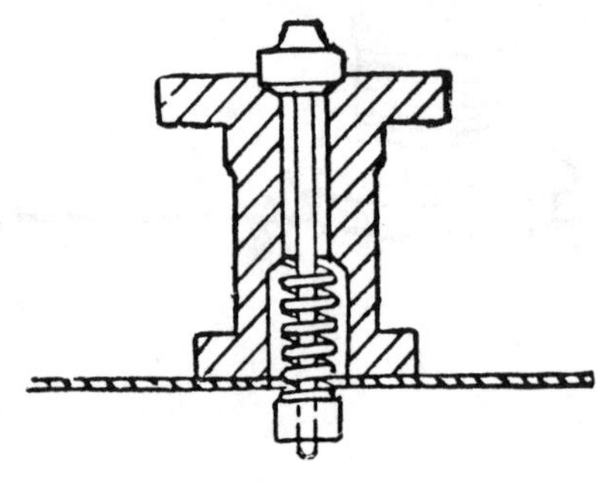

FIG. 13.

tightened up, as required to suit the steam pressure. The spring-balance at its lower end is pivoted to a bracket, which is soldered or screwed into the boiler.

A direct-acting spring-valve (see Fig. 13) is a good one for small models. This can be easily made from the instructions given, or it may be purchased, of any suitable size, from the model-maker's.

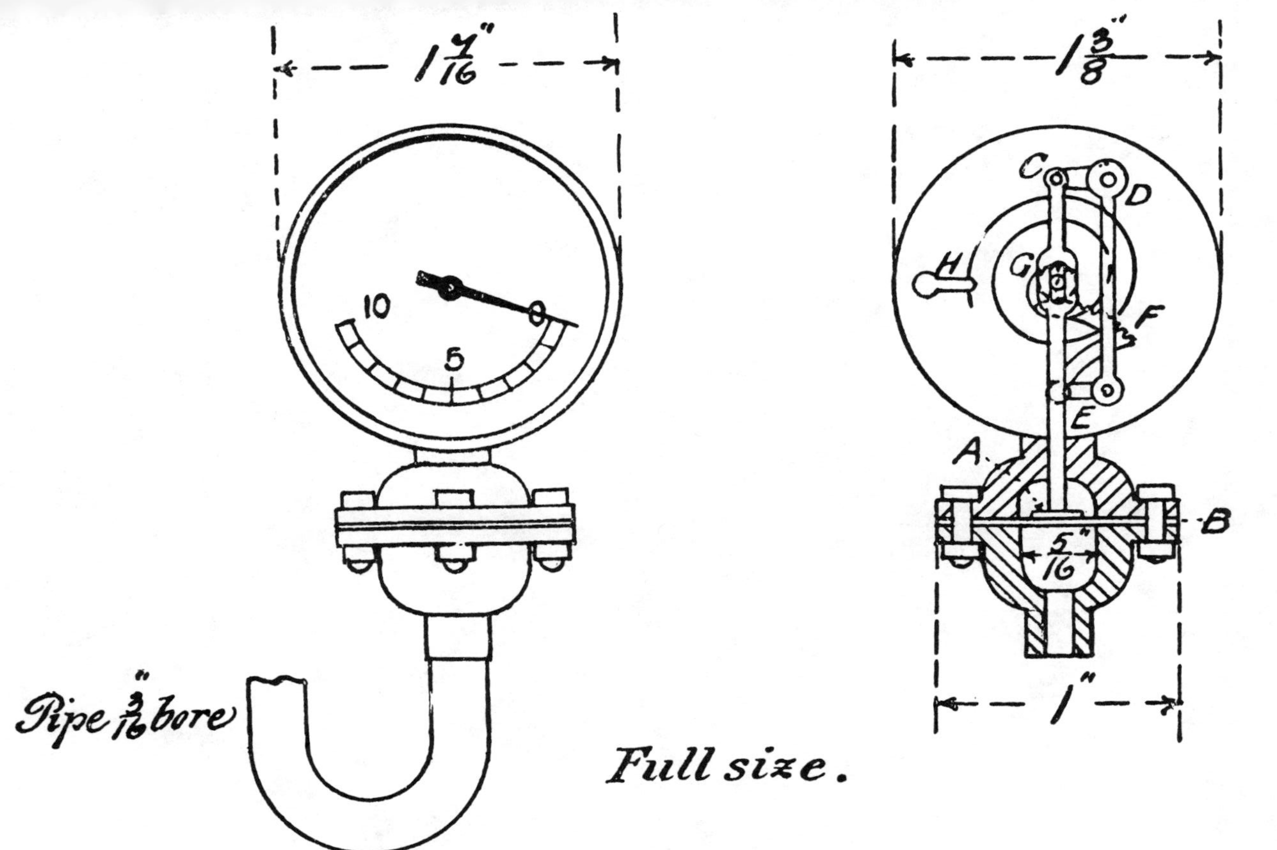

FIGS. 14, 15.

The rule, to find what pressure of steam a valve will hold, is to "divide the length of the lever by the distance from the centre of the valve to the centre of the fulcrum; multiply by the amount of the weight in pounds, and divide by the area of the valve in inches." To be very exact, an allowance must also be made for the weight of the lever and of the valve itself.

To find the area, in square inches, of a circle of a given diameter. Square the diameter, and multiply by ·7854, or multiply 3·1416 by the square of the radius.

A **pressure-gauge** (see Figs. 14 and 15) is an instrument for indicating how much the pressure of steam is above that of the atmosphere. For very small models these are not required, and for small-power boilers it is best to purchase a Bourdon or a Schaffer gauge, as these instruments are accurate, and can be relied upon. However, for those who care to make one, we describe a gauge as made by ourselves, which does well for models. It consists of two short pieces of brass tubing, $\frac{3}{8}''$ diameter, both having a circular collar soldered on to them, where they come together; or this might be a casting, with three small bolts passing through the collars to tighten them up; a thin rubber diaphragm (B) passes between them. The upper piece carries a brass ring into which a dial is fixed, the front being covered over with glass: this forms the gauge proper; the back is a circular plate which fits on to it. The lower tube is screwed on to a bent syphon tube from the boiler. A small flat disc, or piston (A), rests on the rubber diaphragm, the stem of which is flattened and passes up to the top of the gauge (being slotted longitudinally to allow the pointer-stem G to pass through it), and connects with a bell-crank D, which, by means of a lever, pivots on to the end of a drag-link E, attached to a tooth-quadrant F. This quadrant gears inside with a small pinion fixed to the pointer-spindle; a strong watch-spring has its inner end attached to this spindle by a pin, and its outer end to the case of the gauge H; this ensures the pointer returning to zero when the pressure is removed. When the pressure is greater than that of the atmosphere, the rubber bulges upwards, and when it is less it bulges downwards; these motions, being proportionate to the pressure per square inch, are correspondingly indicated on the graduated dial by the pointer. When graduating the gauge, it must be tested by attaching to a large boiler, and marking the dial to agree with that of the boiler as steam is raised. Our gauge is $1\frac{3}{8}''$ in diameter, and indicates up to 10 lbs. per square inch; but by using a stronger spring it might be made to indicate a higher pressure than this. Instead of the rubber, a diaphragm of thin hard sheet brass might be used, to the centre of which a needle-point is fastened.

A **glass water-gauge** (see Fig. 16) is a device for ascertaining the height of the water in the boiler. There are three cocks in the most perfect models: the upper one is fitted into the boiler opposite to the steam space, the middle one opposite to the highest part exposed to the flames, the lower one below the gauge. The upper and middle cocks allow steam and water to pass; these meet in the glass tube, and stand at the same level as in the boiler; the lower cock is to blow out and keep the gauge clean. **To make it**, take two brass blocks, drill a hole through one of them for about half its length, make this hole much wider than the glass tube, fit

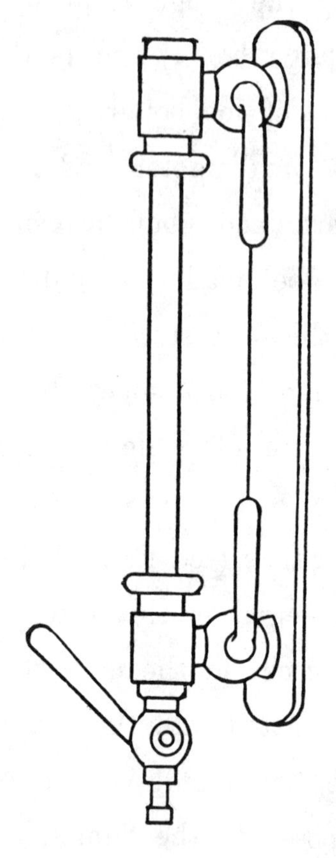

Fig. 16.

a stuffing-box gland on to one end, and screw a cock into the other. Next drill a hole at right angles to the first one, and screw a cock into it; this block forms the bottom of the gauge. Take the other block, fit it just in the same way as this one, except that the hole which is drilled through it for the glass tube to pass should have a plug screwed in at the upper end instead of a cock; drill a hole at right angles as before and screw in a cock. It will now be seen that there are two cocks, one in each block, opening at right angles to the glass tube. Screw both of these cocks into a brass plate, or direct into the boiler. To insert the glass tube, cut it of the proper length, so that when resting in the upper and lower blocks it will not obstruct the passages through the cocks (which are at right angles to it), and lead from the boiler; unscrew the plug from the upper block, insert the glass tube, and push it down till it has entered the lower stuffing-box; screw the plug down on the tube, unscrew the stuffing-box glands above and below, and wind tow round the tube at both ends, where it enters the stuffing-boxes; then screw the glands both up on the packing, and the gauge will be found to be perfectly steam-tight.

A simpler gauge (see Fig. 17), without cocks, is

made as follows. Take two small brass blocks, drill each of them only half-way, or nearly so, to their centre, leaving the hole large enough to admit the glass tube. Now drill two smaller holes in the blocks at right angles to the previous ones, and instead of inserting a cock in each, put in, by

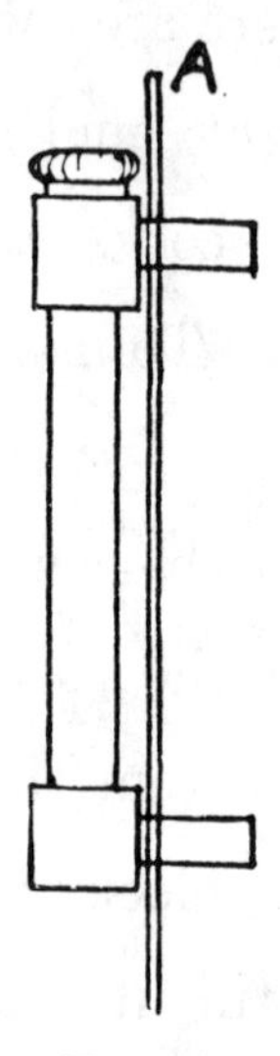

Fig. 17.

screwing or soldering, a bit of brass tubing about $\frac{3}{4}''$ long, the bore of which communicates with the boiler and the glass tube, through the hole in the block previously drilled at right angles to it. Each block is firmly soldered to a small brass plate A, each tube passing through a hole in the plate. The glass tube is inserted by one end into one block, and by the other end into the other block—this is best done previous to soldering down one of the blocks to the plate. When the tube is in position, fill the space between it and the sides of the holes in each block with white lead, and use a wire to see that the communication with the boiler and the glass tube is left quite free. Lay the gauge aside for a

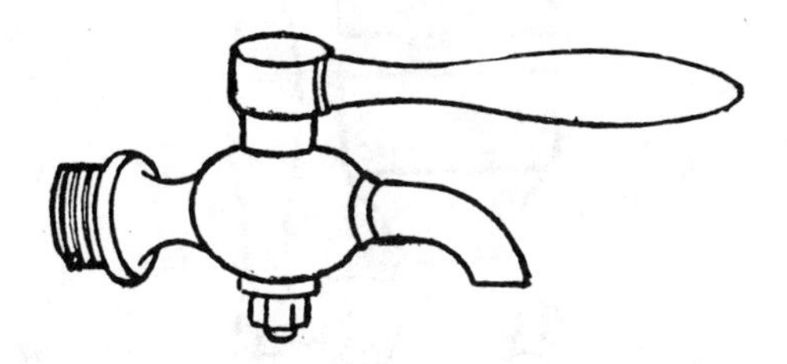

Fig. 18.

week or two near the fire to harden. When this is complete, the gauge will be found perfectly steam-tight, will remain so, and can be attached to the boiler by soldering the plate all round the edges, having previously drilled two holes in the shell, through each of which one of the brass tubes passes when the gauge is fixed on to the boiler.

Gauge-cocks (see Fig. 18). These are fitted to the boiler to ascertain the height of the water inside, and are used instead of a gauge-glass, or in order to check it. For models, two are sufficient; they are to be inserted so that steam should issue from the boiler on opening the upper cock, and water upon opening the lower one. These articles

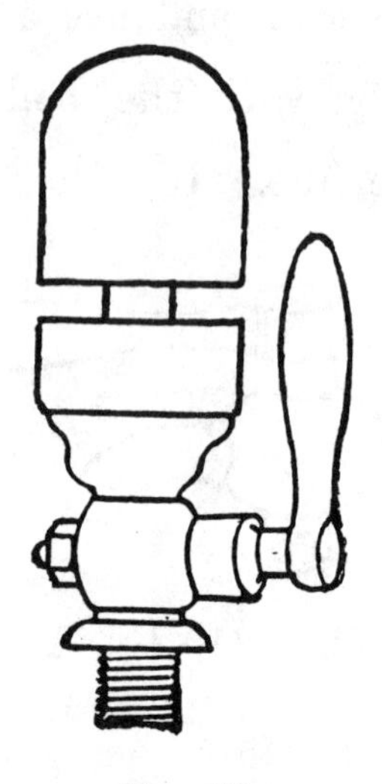

FIG. 19.

are best purchased ready-made, as they are difficult to construct, and can be had from $\frac{1}{16}''$ bore up to $\frac{1}{4}'$ and larger sizes.

The **blow-off cock** is the same as the above, but larger, and is inserted through the shell opposite to the lowest part of the internal fire-box, and is used to blow off mud, scale, etc., from the boiler.

Mud-plugs are not required for small models.

The **whistle** (see Fig. 19) is made out of a brass casting. It consists of a bell-top with a cup-shaped chamber underneath, supported on a hollow stem, containing a cock, which can be screwed into the boiler. The bell-top is screwed on to a small central spindle, which projects upwards, being supported on a disc which has a narrow neck or groove, of a circular shape, cut underneath it. The central hole through the stem must be drilled up to the neck, but no further; then four small holes are to be drilled in a radiating direction from the groove inwards to the central hole. A brass ring is put on, surrounding the disc, leaving a narrow opening between them. This ring is screwed or soldered on to the stem below the groove; the groove, being surrounded entirely by the ring, forms a kind of chamber below the disc, and when steam is turned on, it fills this chamber, and escaping between the disc and ring all round, strikes against the edge of the bell and produces a whistle.

A much simpler whistle is made by taking a bit of brass tubing, blocking up one end, and partially closing the other, by soldering or screwing in a piece of brass rod, with a flat filed on one side. The mouth is made with the corner of a file on the same side as the flat (like an ordinary whistle), just above the bit of brass. The tubing is now soldered on to a cock, and the whole is complete (see Fig. 20).

The **clack-box** (see Fig. 21) is placed just over the

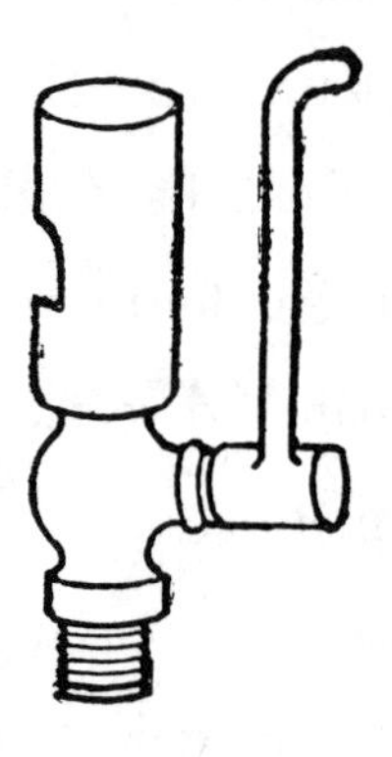

FIG. 20.

top of the fire-box in the delivery-pipe, where it enters the boiler. The water enters at W, forcing up the valve V, and passing through the passage P into the boiler. Thus water passes in, but none gets out, as any backward flow would tend to force the valve more firmly down upon its seat. Only engines provided with a force-pump require this valve. The clack-box is a casting or a brass block. A passage must be made right through the block, of the same size as the delivery-pipe W. From one end take out a second cut from the bore, and stop short of the opposite end, so as to leave a collar between the wide and narrow parts of the bore; this collar forms the seat for the valve V. To the lower end is screwed

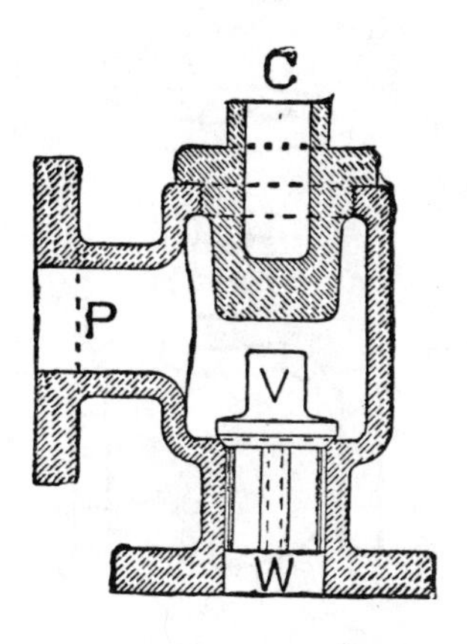

FIG. 21.

a pipe or coupling-screw, which forms a junction with the delivery-pipe from the pump. Into the upper end is screwed a circular brass plug C, which acts as a stop for the clack V, and can be taken off when required to examine the interior. At the centre of the clack-box, on one side, drill a horizontal passage P (of the same bore as the delivery-pipe) which leads into the interior, but higher than the valve V can rise; a short pipe connects the clack-box with the boiler. The valve V is made of brass, or is a casting: it consists of a disc, convex on one side, the other being the edge or face, which is conical. The conical edge of the disc must fit accurately on a corresponding seat formed by the collar mentioned

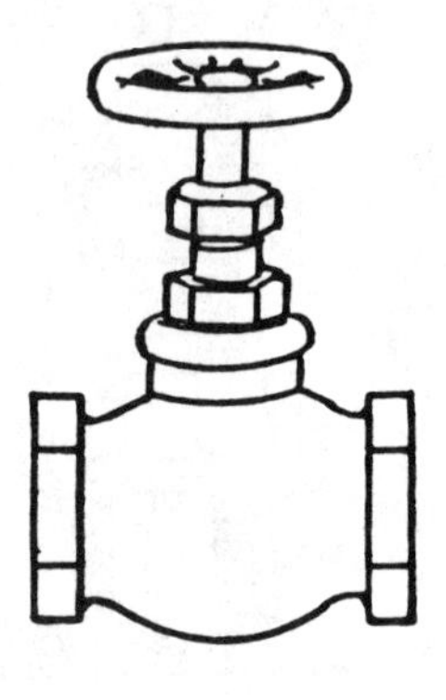

Fig. 22.

above. The valve is guided in rising and falling by three feathers underneath, or by a stem filed flat upon three sides. In large engines, the lift of the valve should not exceed one-fourth of the valve's diameter.

The **regulator.** This in models is simply a cock screwed into the boiler, which carries a coupling-screw or union at its upper end, and is joined by this to the steam-pipe leading to the cylinder.

A wheel-valve can be used instead, and castings for making this may be obtained at the model-maker's (see Fig. 22).

The **blower** is a pipe constricted at the mouth, and let into the funnel, having a cock placed between it and the boiler. When steam is allowed to pass through this pipe, it acts as a blast and intensifies the fire, causing a very rapid production of steam, and is used when the engine is at rest, and steam is low.

The **force-pump** (see Fig. 23). This is used to keep the boiler supplied with water while the engine is running. It is easily made, and gives satisfaction even though small. We give a full-size sectional drawing of a pump suitable for an engine with a $\frac{3}{4}''$ bore cylinder, and it will supply a boiler $10'' \times 5''$.

The **barrel**, **valve-box**, and **plate for screwing to the bed-plate** are all of brass in one casting. It is worked by an eccentric on the crank-shaft of the engine. When cast, reduce the casing to the sizes given, and finish off by boring and turning the barrel and the interior of the valve-box. Make the pump 1″ × ½″ thick, with a ½″ projecting plate on one side; fit a stuffing-box and gland to the mouth of the barrel. Make the plunger out of a piece of brass rod, cut a slot at one end for pivoting to the

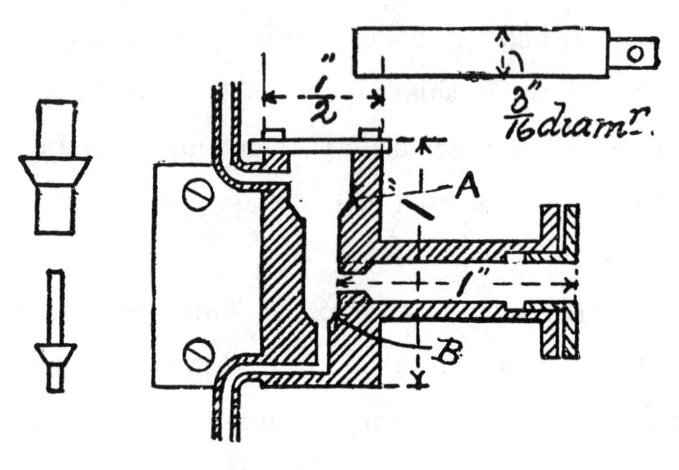

FIG. 23.

eccentric-rod, turn the rest of it circular, to be an easy fit within the barrel; it must be packed with tow in the stuffing-box, and the gland screwed up firmly on the packing. Next, from the upper end of the valve-box drill a hole fully $\frac{5}{16}''$ wide, to a depth of ¼″; bore a second hole, $\frac{3}{16}''$ diameter, to a depth of $\frac{1}{2}''$; and lastly drill a third hole, $\frac{3}{32}''$ bore, to the depth of $\frac{1}{4}''$. These holes being all continued in one vertical line, entrance holes for the delivery- and suction-pipes must be drilled into them at the top and bottom of the pump. The taper, which it will be noticed is given for a short distance to each of the holes by the different-sized drills, will do for the valve-seats. It will be seen that the delivery-pipe enters just above the valve-seat A, and the suction-pipe below the valve-seat B; the barrel must be bored through right into the central aperture. The valves are two in number, made of brass wire and turned on the lathe. The larger is the delivery-valve, nearly $\frac{5}{16}''$ across the face; the upper stem is the thickest, and is left circular; the lower one is smaller, and of a triangular shape. A hole must be bored right up the centre of the stem of the large valve to a depth of $\frac{1}{4}''$; into this hole the spindle of the suction-valve passes, when both valves are in their places, and so the larger valve regulates the lift of the smaller one. The suction-valve is the lower, and is much smaller than the other, the facing being nearly $\frac{3}{16}''$ diameter. The valves must be made tight according to the method already described: the upper opening of the valve-box should be closed by a flat bit of brass, made air-tight, and held in position by screws. The lift of the upper valve is limited, by striking against this piece of brass, and depends on the length of the stem above the valve face. To insure success, make the delivery- and suction-pipes as short as possible, or in other words, have both tank and pump as near the boiler as convenient.

Action.—The plunger works in and out of the barrel, leaving a space behind it when drawn out, which is supplied with water through the suction-pipe and lower valve, and this water, not being able to get past the lower valve, when the plunger is pushed inwards, must force its way past the delivery-valve, and through the delivery-pipe to the boiler. A check-valve ought to be placed at the mouth of the delivery-pipe, where it enters the boiler.

The **injector** (see Fig. 24) is an instrument for supplying a boiler with water, and is used instead of a pump. It is made from a casting, and is very difficult to fit up. A small model does not work well; there is difficulty with the nozzles, and the steam is of a low pressure: it does best with high-

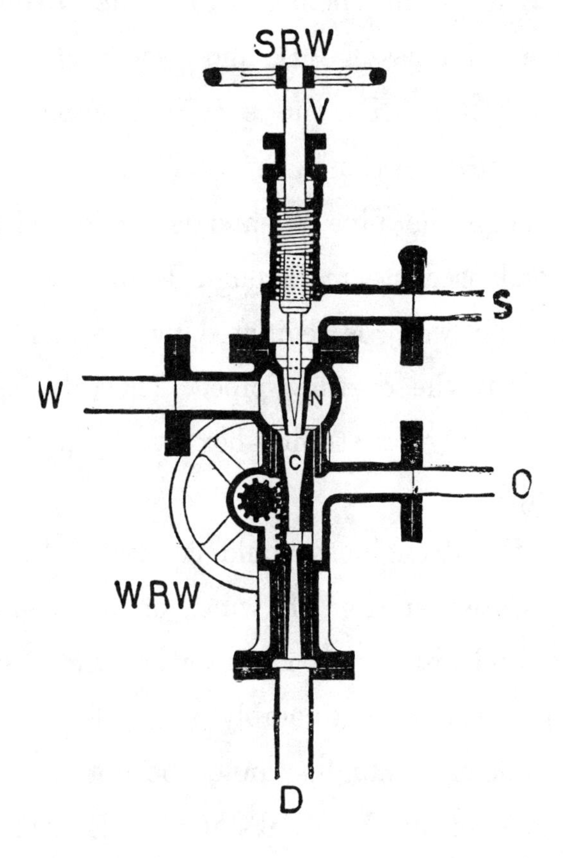

FIG. 24.

pressure steam. For these reasons we pass it by, except giving a sectional drawing of one, and a short description of how it acts. Steam from the boiler is admitted by pipe S into the injector and passes through the nozzle N; this passage is opened and shut, and the amount of steam admitted through it controlled, by the spindle SRW; the steam then runs along the passage D, and past a check-valve into the boiler. By the suctional action of the current of steam, water is drawn in from the water-tank through the pipe W, and is carried with the partly-condensed steam through D into the boiler. The supply of water is regulated by the hand-wheel WRW; O is the overflow-pipe. The valve-spindle V passes through a stuffing-box at the upper part of the injector.

Fuel. Small engines should be heated by means of methylated spirits of wine, or a gas-burner. Charcoal and ordinary coal are out of the question, unless the boiler is tolerably large, for in small boilers sufficient draught cannot be got to support the combustion of these substances. If our readers must have solid fuel for their boilers, we may state that blocks of wood cut into sizes $1'' \times \frac{1}{2}''$, or less, and steeped in methylated spirits, will burn in a very small boiler, provided there is a grate; but it requires constant stoking to keep the fire up. We have seen a brazed vertical copper boiler $12'' \times 7\frac{1}{4}''$, with an outer fire-box $7\frac{1}{4}'' \times 6''$, which burned coal, and drove an engine having. a $1''$ bore cylinder, and did fairly well. The flue-tube was $2\frac{3}{4}''$ diameter, and about $18''$ long. The grate was made of a circular hoop of sheet-brass fitting inside the fire-box, across which was riveted half-a-dozen $\frac{3}{16}''$ iron fire-bars. The whole thing rested upon brackets inside the fire-box.

CHAPTER III.

THE STEAM-ENGINE.

THE **steam-engine**, in whatever form it exists, consists of a bed-plate, which carries two distinct mechanisms: (1) The driving mechanism, viz. the piston, piston-rod, cross-head, connecting-rod, crank, crank-shaft, and fly-wheel; (2) the valve-gear, viz. the slide-valve and spindle, and the eccentric, with its rod and strap: these control the admission of steam to one side of the piston, and the exhaust from the other. In addition to these there may be other mechanisms added, viz. a governor, to regulate the speed, a link motion to reverse the engine, and a separate cut-off valve, etc.

The Parts of a Steam-Engine.

The **bed-plate** is the frame on which the engine rests, and is either a brass or iron casting, containing the water-tank, or a plate fixed on a wooden block, or supported upon turned pillars. It should be filed, or planed quite flat upon the top, before centre lines are scribed upon it, for getting the parts of the engine into line with each other.

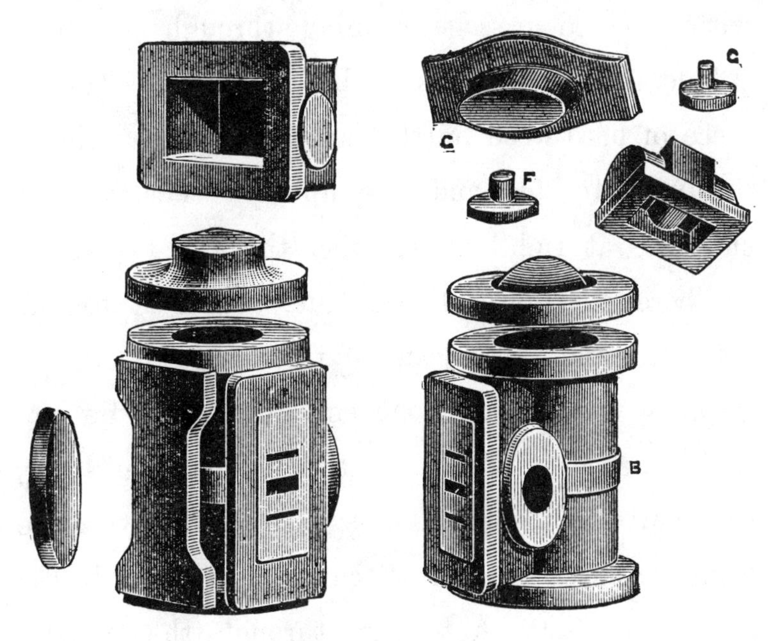

FIG. 25.

The **cylinder** (see Fig. 25) is the most important part of the engine, and unless accurately made, the engine will turn out a failure under steam. It is a brass or iron casting (brass is the easiest to work), and unless bored by the maker when ordering—

which can be done for an extra shilling or two, and will save much trouble—must be done by the amateur himself. This is a difficult task, and requires careful workmanship to insure accuracy. The cylinder, instead of being cast solid, is provided with a cored passage running through it, about $\frac{1}{16}''$ less than the finished bore is to be. Turn a piece of hard wood on the lathe, drive the cylinder firmly on to this, and face up both ends, so that they are at right angles with the cored passage. To bore out the cylinder true, attach it to the lathe face-plate by means of three dogs or clamps, gripping the flange of one end all round. Be sure that, when the cylinder is chucked in the lathe, the centre of the cored passage is in line with the lathe-centres. Get a steel cutter made by a blacksmith, long enough to pass through the cylinder from end to end, and having four parallel cutting edges large enough to take out a $\frac{1}{32}''$ cut. The point of the cutter only should have a slight taper, just sufficient to enter the cored passage, and the tool should be tempered so hard, that the cutting edges will remain sharp all the time while the cylinder is being bored out. Insert the cutter by one end into the movable lathe head-stock, and by the other into the cored passage in the cylinder. After chucking, feed the cutter forwards by means of the hand-wheel, while the cylinder is revolving in the lathe; the cutter is kept from revolving by setting the T-rest at right angles to the lathe-bed, and letting the tool slide along the top of the T-rest, as it advances forward into the cylinder. Do not stop the boring process, till the cutter has gone right through the cylinder, after which the bore will be found quite true and perfectly circular, if the cylinder has been properly chucked. A second cut can be taken by means of a rose-cutter (slightly tapering at the point only), while the cylinder is still chucked; and after this has gone through, the cylinder will be found perfectly true and as smooth as glass inside. This second cut is not absolutely necessary, if the steel cutter has been well made, with the four cutting edges perfectly parallel to each other, and of a temper hardened, so as to remain quite sharp during the whole of the drilling process. A little oil can be used to facilitate boring operations. The cylinder is now smoothed all over with a file, and can be painted where it has not been turned. It will be noticed that as yet the steam- and exhaust-ports have not been bored out.

Steam-ports (see Fig. 26). In large cylinders the steam-ways are cast in; in small models they must be drilled. Take the cylinder, square up the valve-

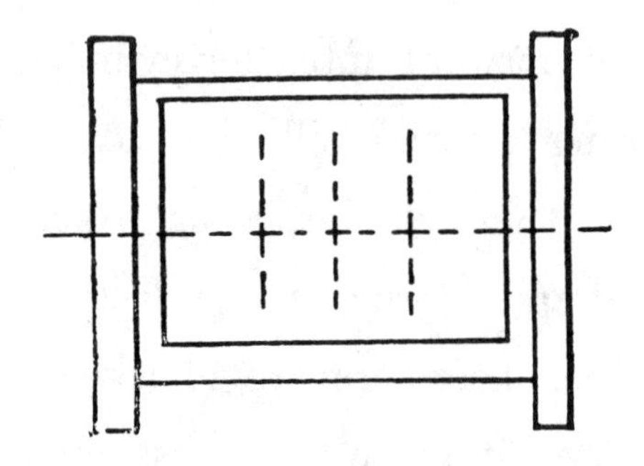

Lines scribed on cylinder valve-face, where they intersect parts, must be drilled.

FIG. 26.

face, and flatten it by filing; measure off and draw a vertical line through the centre. Now draw two vertical lines, one being on either side, each distant about two and a half times the width of the finished steam-port from this central line. Next draw a horizontal centre line, where these lines intersect each other; punch three holes to form the ports—the middle one is for the exhaust, the outer ones for the admission of steam. Next mark the steam-ports on the face of each flange, not far from the bore; punch these holes, which must be drilled in the middle line, of the thickness of the ridge that runs from the valve-face to the cylinder mouth. Fix the cylinder in the vice, drill the exhaust-port for a short distance inwards, but stop short of piercing the bore; drill a hole to meet this one at right angles: this forms the exhaust-port. With a smaller-sized drill bore the steam-ways straight inwards for a short distance; after this is done, with the same drill drill a hole from either end into each steam-way, and be careful not to run through them, and let the mouth of the drill enter the exhaust-port. Now with a small chisel square the ports, and prolong them vertically for a short distance on either side of the horizontal centre line. This makes the engine steam better. At each end cut a slot out, so as to make a steam-way between the ends of the port-holes and the cylinder bore, that they may remain open for the passage of steam, when the covers are bolted on.

In a $\frac{1}{2}''$ bore cylinder the steam-ways can be drilled with a $\frac{1}{16}''$ drill, and the exhaust-port with a $\frac{3}{32}''$ drill, as the exhaust-port must always be wider than the steam-ports.

The **covers.** These are simply circular discs of brass or iron which close up the ends of the cylinder. The front one contains a stuffing-box, and has the piston-rod passing through it; but in all other respects it is similar to the back cover. It is best to fit up the **back cover** first, as it is the easiest. Find the centres, and centre punch on both sides; take a long, narrow strip of sheet brass, drill a big hole in one end to allow of the lathe-centre entering the central hole in the cover when chucked, solder this strip on to what when finished will be the outside, chuck in the lathe, and this brass strip will rest against the crank of the mandrel, and act as a lathe-carrier, taking the cover round with it. Turn the edges to the same diameter as the cylinder flanges; face it up truly. Some covers have a projecting part which enters the cylinder bore when in position, but this is not required. Now remove the cover, undo the solder, and reverse the cover in the lathe. Having soldered the strip of brass to the previously-turned face, turn this side as before, remove the strip, file off all the solder, and the cover is finished. A hole can be drilled through the centre, to screw in a lubricator (see Fig. 27).

The **front cover** (see Fig. 28) is fitted up in the

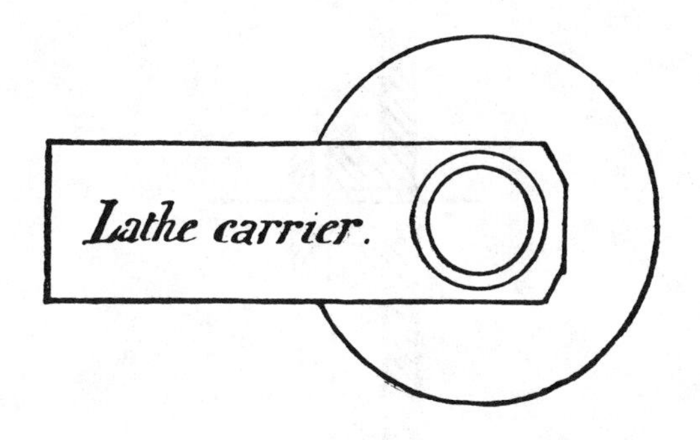

FIG. 27.

same way. Drill a hole right through it, so as to be an easy fit on the piston-rod; widen this hole for about half its depth, beginning at the outside or stuffing-box end. With a plug-tap make a screw in it. If the hole for the piston-rod is $\frac{1}{16}''$ diameter, that for the stuffing-box should be $\frac{1}{4}''$ diameter, continued for a depth of $\frac{5}{16}''$. Take a piece of brass $\frac{7}{16}''$ long and $\frac{3}{8}''$ diameter, bore a central hole $\frac{1}{16}''$ diameter in it, turn a piece of iron wire so as to fit this hole, fix the brass on it by soldering at one end, chuck in the lathe, turn a projecting flange at one end, and then turn down the rest, till it will tap with the same thread as was used for the stuffing-box; remove the piece of brass from the iron wire—this forms the

Fig. 28.

stuffing-box gland, and can be screwed into place. Sometimes, instead of the gland being screwed in, it is held in position by two bolts and nuts, in which case, the stuffing-box is cast on the cylinder cover of an oval shape. The flange of the gland is also oval, and instead of being screwed up against the packing, the neck simply fits closely inside the stuffing-box, and is held up against the packing by two bolts passing through its flange, which have nuts to keep it in place. The bolts are screwed into the cylinder covers. This plan is not a good one for small models, as it is very troublesome to remove the glands for re-packing.

To fasten on the covers. Drill four bolt-holes through each cover near the circumference, equally distant from the centre, and at the same distance from each other; fix the cover in its place upon the end of the cylinder (a little solder does well to hold it, and this can be filed off afterwards), mark upon the flange the position of one of the bolt-holes, insert a drill through this hole in the cover, and drill through or partly through the cylinder flange, tap a screw-thread, and insert a screw through the cover and tighten up (its head holding the cover on). Next, with the same drill bore three more holes in the flange, corresponding to those in the cover, while this latter is still in position, and in this way the holes must be correct, and correspond with the holes drilled through the cover; put in three more screws, and tighten up. If a little red lead is put in between the faces, this will make the joints quite steam-tight. A flat india-rubber ring put in between the cylinder flange and cover, if the latter is tightened up on it, keeps the parts steam-tight; but a bit must be cut out of the ring previously, which corresponds to the position of the steam-port, to leave this uncovered for admission of steam to the cylinder. This plan does for large models. Small bolts may be screwed into the cylinder flanges, the covers made to slip over them, and be tightened up against the cylinder with nuts. Bolts and nuts are used in large engines for holding the covers on.

The front cover is fastened on to the cylinder in the same way, but before drilling holes through the cylinder flange to correspond with those in the cover, the piston must be finished, packed with tow, and placed in the cylinder, with its rod passing through the gland, screwed up tight in the stuffing-box. Push the cover up against the cylinder face, when it will be found that the stuffing-box is in line with the bore; and if corresponding holes be now drilled in the cylinder face and the cover bolted on, the piston will work quite easily backwards and forwards, if not packed too tight—a little oil is a great help to lessen friction.—

The **steam-chest** is a separate casting, flat on the outside, with a projecting part which, when bored through, will form the steam inlet. The inside is hollow, and is surrounded by a square flange, which is bolted to the valve-face; the front has a projecting part also, which is to form the slide-valve stuffing-box. The inlets for the steam and the stuffing-box only require turning; the other parts are to be filed up smooth and square. If the inside is very rough, it may be smoothed down by taking pieces out with a small chisel (when gripped in the vice), so as to give plenty of room for the travel of the slide-valve. Fit up the stuffing-box and gland as already mentioned. Drill the inlet for the steam, and tap the hole for screwing in the steam-pipe. Drill four bolt-holes through the chest near each corner, and widen these with a rimer; place the chest in position against the valve-face of the cylinder, drill a hole at each corner, and attach the cover with bolts and nuts, following the directions given for the cylinder covers. Be careful that the chest is attached in such a way that, when the piston-rod and valve-rod are in position, they remain parallel with each other. The chest is held in position by four nuts, and is made steam-tight by red lead or india-rubber.

Where the exhaust-port opens out above the cylinder, the mouth should be widened with a rimer, to a depth sufficient to take a screw-thread; it may then be tapped, and the exhaust steam-pipe screwed in, which ought to be of a rather larger bore than the steam-pipe. It is a good plan to screw a lubricator and cock into the upper part of the slide-valve case, in order to lubricate the slide-valve, for if well oiled it works more easily.

Pet-cocks are employed at each end of the cylinder to let out the water that condenses, when steam is admitted to a cold or partly-cooled cylinder. They are screwed into two lugs, which are cast upon the inside of the flanges, it may be on the under surface. Holes must be drilled through into the bore of the cylinder before screwing in the cocks; both cocks may be connected together by a rod, so that they can act together. In small models these are unnecessary, except for the sake of appearance.

Horizontal cylinders are fastened to the bed-plate by means of bolts passing through feet or lugs cast upon them (see sketch), or by bolting them to the front covers, which are cast on the bed-plate. Vertical cylinders are bolted through the covers to the top or bottom of the frames that support them.

A **lagged cylinder** is one with strips of wood resting on the flanges, fitted together and held in place by two metal bands, leaving a space around the cylinder which may be filled in with felt, or some non-conducting material, as hair. Lagging need not be applied to small cylinders, as it is troublesome to fit on properly. Its use is to prevent radiation, and render the cylinder hotter.

The **piston.** In small engines this is simply a circular brass disc, with a groove to hold the packing. The piston must be turned quite circular, be an exact, yet an easy fit in the cylinder. It is screwed to one end of the rod which passes through its centre. For small engines it is impossible to turn the piston upon the rod in the lathe, as we do when making the larger sizes; so it is best to drill the piston through the centre, fasten it by solder upon a temporary shaft (of iron wire) fixed between the lathe-centres, and after turning, it will be found to be perfectly circular, and the hole passing through it coinciding with the centre. A piston suitable for a $\frac{3}{8}''$ or $\frac{1}{2}''$ bore cylinder should be made $\frac{1}{4}''$ broad, having a $\frac{1}{8}''$ groove for packing, with a $\frac{1}{16}''$ collar on each side of it. If these pistons are properly packed at first, they will give very little trouble afterwards, as the collars prevent the packing from being blown off by the steam. The narrow pistons found in cylinders purchased in the shops, with a $\frac{1}{16}''$ groove and a narrow ($\frac{1}{32}''$) collar on each side, after being run by steam once or twice become useless, as the packing gets blown off if the pressure is at all strong. The piston is made steam-tight by packing with hemp or tow, well saturated with oil or melted tallow. After packing, work the piston backwards and forwards by hand in the cylinder till it goes easily. Pistons must never be made a tight fit.

In large models and small-power engines, the piston is an iron casting composed of two separate pieces: the piston proper, and a circular disc which goes behind it, the "piston follower." Turn the piston-rod out of a piece of steel or iron wire, centre both the piston and follower, centre punch and drill through, screw them both on to one end of the piston-rod (this part of the rod being turned down narrower where it enters the piston, so as to leave a collar against which the piston can be tightened up). A little piece of the rod should project at the back through the piston for a nut to tighten the piston up against the collar; chuck the piston-rod and piston in the lathe, then turn and face the piston while revolving on the piston-rod; turn the piston just a shade less than the bore of the cylinder. To render this size of piston steam-tight, one, or it may be two, brass or iron rings are made to surround it called "piston-rings," and to get them over the piston the follower must be removed from the piston-rod. In large engines the piston-rod is tapered where it passes through the piston.

Piston-rings (see Fig. 29). The ring is an iron or

brass casting. File the inside of the ring with a round file, to fit the groove on the piston. In order to turn the ring, turn a piece of hard wood and drive the ring firmly on to it with a mallet, chuck this in the lathe, and turn the ring to a size a little larger than the cylinder bore; face the ends. Next put the ring in the vice, gripping it at the sides with a fine drill, drill three or four holes through it

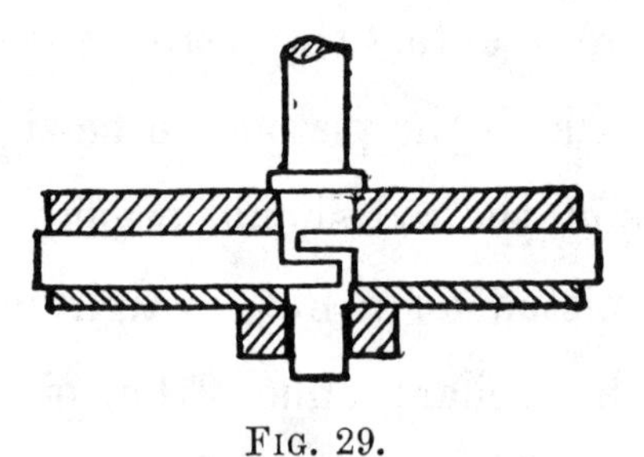

FIG. 29.

lengthways in the circumference, close to each other. With the frame-saw make a cut from the edge right through to the first hole; do the same from the opposite side to the last hole in the row, give a tap with a chisel, and the ring will split open; file the edges of the cut smooth, and bring them close together by filing, so that by compressing the ring tightly with the hand it will enter the cylinder. Place the piston in the cylinder (having removed the follower), and by compressing the ring it will slip over the piston; place the follower in position, tighten up the nut, and it is finished. The spring of the ring against the cylinder keeps the piston steam-tight. The elasticity can be increased by hammering the ring all round upon the inside with a light hammer, while it is resting on a block of wood, before placing it on the piston. One ring will do for models. Making the "split" according to the above method causes less escape of steam than if the ring be simply sawn across. When two rings are used, they are placed upon the piston so that the split occupies different sides.

The **piston-rod** is made of steel or iron wire turned on the lathe, and of such a length that when the piston is close up to the back of the cylinder, or on the "inner dead centre," the cross-head keeps clear of the stuffing-box. Their diameters range from one-sixth to one-tenth the diameter of the cylinder bore in actual practice. The rod is made steam-tight where it passes through the stuffing-box by means of tow or hemp, and the gland is tightened up on the packing as required (see Fig. 29).

The **cross-head** is a cast brass or iron block, screwed to the end of the piston-rod, and forms the junction between the piston-rod and the connecting-rod, by means of the cross-head pin, which is screwed into it. The head is either slotted out at the end to take the connecting-rod, or the latter is forked, in which case the head is solid. Projecting pieces are cast on the top and bottom, which must be smoothed; these work on the guide-bars, and thus guide the motion of the piston-rod. These guides, instead of being cast on, may be separate, simply flat blocks of brass with a shoulder inside, and into the centre of which the cross-head pin passes, after passing through the cross-head, and these "guides" or "slippers" slide along between two parallel guide-bars on each side, one above, and one below.

The **guide-bars** should be an easy sliding fit for the cross-head "slippers." They must be set truly in line with the axis of the cylinder bore, so as to guide the piston-rod when travelling in a straight line. They must be filed smooth all over, and the guide surfaces made quite flat. The lower guide bars are bolted at each end to the bed-plate; the upper bars are bolted to the lower ones, washers, or distance-pieces, being inserted between them. A small hole is drilled in the centre of the upper bars for oil, or an oil-cup can be screwed in. Instead of this, there may be four brass pillars, each having a lug for attaching the bars to; these pillars are bolted to the bed-plate, two on either side of the piston-rod. The bars are made out of a piece of iron filed flat, each cut to the same length, drilled at both ends, and fastened by bolts, passing through the lugs in the pillars. In this case, the "slippers" must be filed down, so as to be an easy fit between the bars. The pillars must be fixed to the bed-plate on each side, so that they keep clear of the cross-head guides, when the piston is on the dead centres, or in other words, that the guiding surface must be longer than the stroke.

In vertical engines the cross-head simply slides up and down between the frames, being kept in place by a groove, cast upon opposite sides of the guide-frames.

The **connecting-rod** (see Fig. 30) is made of soft iron, cut to the proper length, circular or square, but having the ends flattened, the crank end being the larger of the two, and in its simplest form is just this, with the addition of a hole drilled through at each end, one for the cross-head pin, and the other for the crank-pin; but so simple a rod as this is not very "engine-like," and besides, with a centre crank, it could not be got over the crank-pin. On the

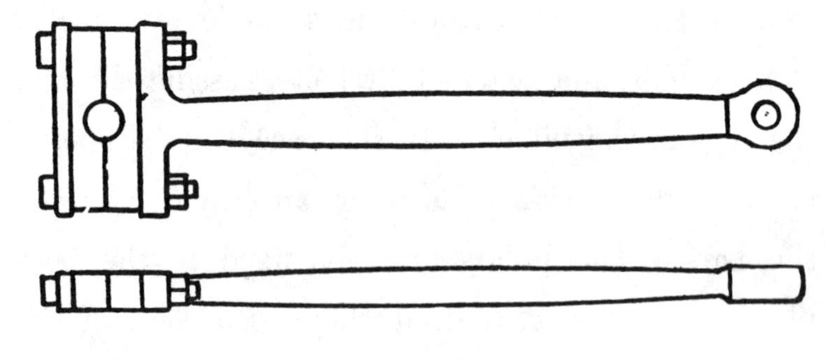

FIG. 30.

other hand, a correctly-modelled connecting-rod, brass bushed, with jib and cotter for tightening up the brasses, would be very difficult for an amateur to construct in such a small size, so we describe and illustrate a form of rod of the "marine type," which is tolerably correct, and if well finished has a good appearance. Get a blacksmith to forge a T-shaped piece of iron, centre punch, and turn it on the lathe, tapering slightly towards each end. The cross-head end is solid; file it circular at the end, and bore it out for the pin; or this end may be forked. Drill two bolt-holes through the T-shaped head, at equal distances from the centre, smooth and flatten a piece of brass, of the same thicknes as the end, drill two holes through this to correspond with those in the T-head, put on a flat piece of iron at the end—this gives a better finish. With a frame-saw, saw the brass longitudinally through the centre, take off the rough cuts with a file, make two bolts and nuts, pass the bolts through the brass and the strip of iron in front of it, and bolt on to the T-head. This end of the connecting-rod is finished, and contains the "brasses," and is ready to be drilled for the crank-pin. Centre, and punch a hole, then drill so as to make the hole an easy fit on the crank-pin, and it will be found that half of the hole just drilled is in one brass, and half in the other. After polishing, remove the bolts, slip the crank-pin through the brasses, tighten them up, and the connecting-rod is complete. There are no means of adjusting the brasses for wear, but this is practically nil in a model engine.

The connecting-rod is usually made six times the length of the crank, or it may be made two and a half times the length of the stroke.

Bearings and **plummer-block** (see Fig. 31). This is a casting, having a sole for bolting to the bed-plate.

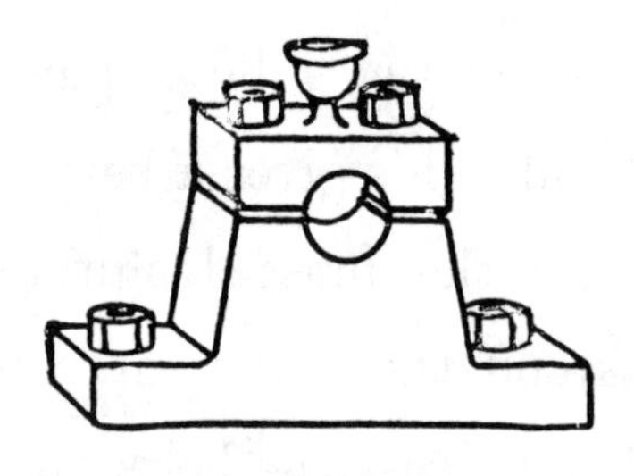

FIG. 31.

File it all over, and make the under surface of the sole that rests on the bed-plate perfectly flat and true, or else it will rock when the engine runs. Drill bolt-holes into each end of the sole, widen these with a rimer, so that they will pass easily over the bolts, screwed into corresponding holes in the bed-plate; when fixed, the plummer-block is retained on the bolts by nuts. From the top of the bearings drill two holes (near the ends) downwards, about half the depth of the plummer-blocks, bolt down both bearings to the bed-plate (having previously bolted down the cylinder), mark on the outside of each a horizontal line, which exactly corresponds to the horizontal central axis of the cylinder (this is done to get the engine fitted truly "in line"). Make a saw-cut horizontally through this line in both bearings, the detached piece forms the cap, or upper half of the "brasses"; widen the vertical holes in this with a rimer, so that they will easily slip over the bolts, screwed into the plummer-block, and be tightened up by nuts. Mark the centre of the saw-cut in both. Drill a transverse hole through each, of such a size as allows the journals of the axle room to revolve freely inside, then drill a small hole through the cap for oil.

Note.—Bearings should be fixed to the bed-plate, so that the crank-shaft is at right angles to the piston-rod, and exactly in line with it, when the cylinder is bolted down in position.

Standards for vertical engines are fitted up in the same way, and must be so bolted to the bed-plate, that the central axis of the crank-shaft will stand vertically over the centres of the piston-rod and eccentric-rod. In an ordinary steam-engine, each plummer-block is an iron casting, having a sole, through which pass the holding-down bolts;

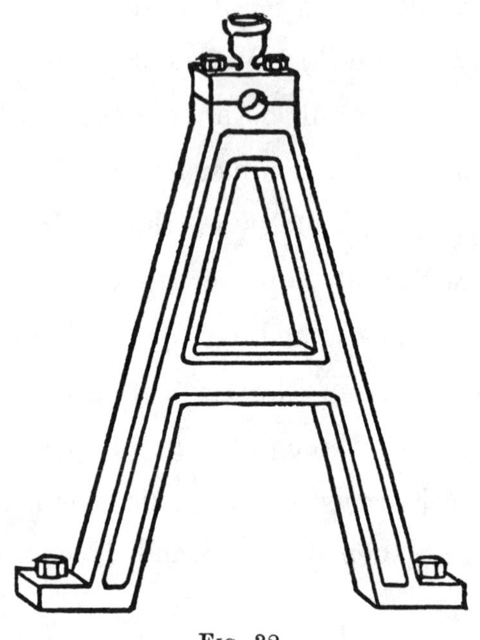

FIG. 32.

the cap is also a casting. Between the block and the cap is fitted the brass bush to take the axle. This bush is in two halves, which are called the "brasses" or "steps"; these are cast with flanges on both sides to keep them in position, and to prevent them turning along with the axle. They are furnished with lugs, which enter corresponding recesses in the block and cover (see Fig. 32).

The **crank-axle** (see Fig. 33), by means of the crank, converts the reciprocating, to-and-fro motion of the piston into a circular motion, through the intervention of the piston-rod and connecting-rod. When the crank is set at one end of the shaft, it is called a "single crank." When there is a

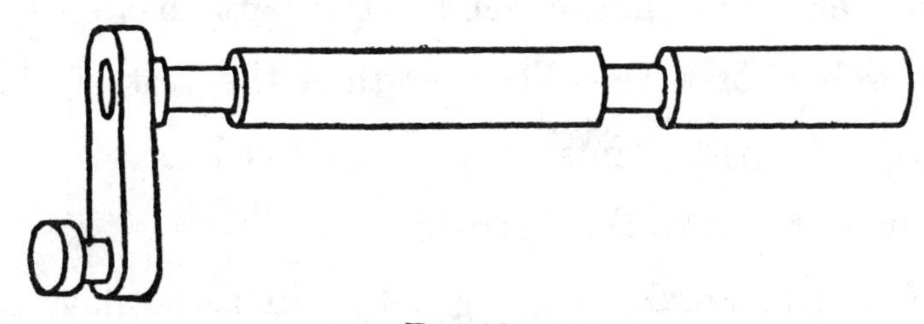

FIG. 33.

bearing on both sides of the crank-webs, it is called a "centre crank."

Note.—Whatever form is adopted, the crank-web, or distance from the centre of the axle to the centre of the crank-pin, must exactly equal half the distance of the stroke.

The **single crank** is easiest to make. Take a piece of iron wire (the proportion of the shaft is ·33 of the diameter of the cylinder), cut to the proper length, chuck and turn in the lathe, turn down to a smaller diameter, the parts which form the journals, then screw the shaft to the crank-web.

To make the web, take a piece of brass, at the proper distances to suit the length of stroke, drill two parallel holes straight through the web, screw the end of the shaft tightly into the largest hole, and into the other hole screw the crank-pin, till the shoulder bears on the web (as the part that enters the web is of a less diameter than the rest of the pin). All the joints may be soldered as well as screwed, to make them strong.

A **centre crank** (see Fig. 34). In these days of cheap castings and forgings, the best plan perhaps will be to buy a solid, unfinished bent steel crank-shaft, as sold by the model-makers. This can be turned and finished on the lathe, and when finished will prove far stronger than a built-up crank-shaft. They are sold in sizes ranging from a $\frac{1}{2}''$ throw up to small-power sizes.

A **built-up crank-shaft** is made as follows. We assume that the axle has been turned, accurately measured, cut to the proper length, the journals and crank-collar turned. Take two flat pieces of brass (for the webs), make them of the same length, mark and centre punch holes at the distances to give the proper throw. Be sure the holes are drilled straight through each web; the widest hole at one end is for the crank-shaft to pass through, the narrow hole at the other end is for the crank-pin. Take a piece of wire for the crank-pin, file down a bit at both ends,

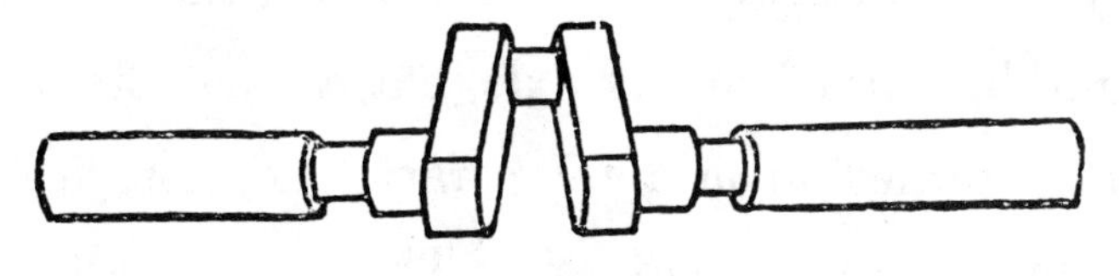

Fig. 34.

leaving a shoulder, screw both shoulders firmly into the webs, and if the holes have all been carefully bored and centred, the axle will run perfectly true, when placed in the lathe. The crank-pin does not require turning, but the corners can be taken off the crank-webs with a file. Solder the parts together.

The **fly-wheel** is a brass or iron casting, containing five or six spokes. Take the wheel, centre it accurately upon both sides, and punch the centre holes; then drill half-way through from both sides of the central boss; the hole must be a sliding fit on the crank-axle. Turn a piece of wire, to be a tight fit in the wheel; with a mallet drive the wheel on to it, attach with solder if required, and chuck in the lathe; turn and face the wheel. The diameter is generally made about four times the stroke. Remove the wheel from the lathe, and file the inside of the rim and spokes smooth; these can be painted afterwards. In the lathe, the fly-wheel might be turned on its axle, but a small fly-wheel is best turned upon a temporary axle, from fear of bending the axle in any way. The wheel is attached to the axle by screwing, soldering, or keying on. In this last case, the shaft must be filed away a little on one side, so as to be flat over the fly-wheel seat; with a fine file, a small recess or key-way must be cut in the boss of the wheel. When placed in position, this recess in the wheel must correspond to the flat or key-bed on the shaft; a small wedge-shaped key is driven into the space between the wheel and the shaft, and the wheel will now revolve quite true with the axle.

The **pulley-wheel** is a brass or iron casting, fitted up in exactly the same way as the fly-wheel. It is placed upon the opposite end of the axle to the fly-wheel; sometimes it is provided with a groove.

The **slide-valve** (see Fig. 35) is a brass or iron casting, which must first be filed up square and smooth all over; then with a small steel chisel—

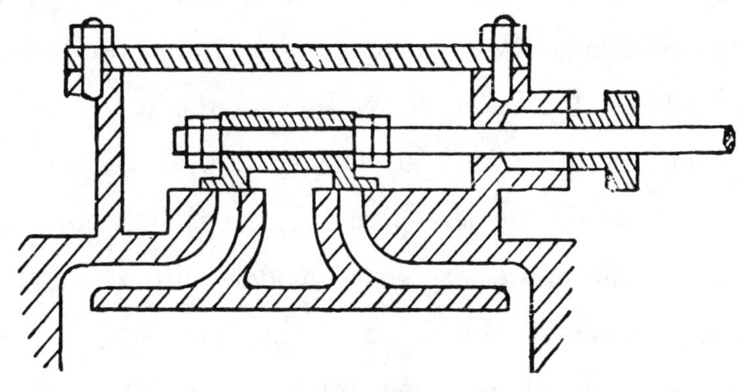

FIG. 35.

while the valve is gripped in the vice—cut or hollow out the interior, until it becomes box-shaped, and having edges all round. The length on the face, from the inside of one flange to the inside of the other, should be just a shade less than the distance between the inner edges of the steam-ports, and the flanges must both be left of the same thickness at the ends. When finished, the length of the valve should be such that when it is placed on the cylinder face in "mid-position," *i.e.* equally covering both steam-ports, its edges will extend a very little way beyond the ports at each end. The amount which the valve projects over the steam-port on the outside is called the **outside lap** of the valve, and the amount which it projects on the inside is called the **inside lap.** We may disregard this latter term, and when "lap" is used, we mean outside lap only. By means of "lap" an earlier cut-off of steam is obtained during the piston stroke than would be got without it. (Very small models will work well enough without the valve having any lap.) The valve-face must be filed (or scraped) quite flat. This is best done by rubbing the valve, with its face downwards, upon a flat file, laid upon the bench; do the same with the cylinder face, then grind both of them on the face-plate with emery powder and oil. Lastly, do the same, grinding the one upon the other: when completed, the valve ought to be quite tight on the chest. To test, wet the valve and cylinder face with water or oil, place the valve in mid-position (holding it tightly up against the cylinder face), and attach a piece of rubber-tubing to the blast-pipe soldered into the exhaust outlet. On blowing through the tube, no air should pass, or bubble out at the sides of the valve. Block up the inlet and outlet of the cylinder bore, move the valve, so that the back steam-port and exhaust-port communicate alternately; then, on blowing through the exhaust-pipe in either of these positions, if the valve is held up tightly, no air passes out at the sides. The valve is secured to the valve-spindle by nuts screwed on the spindle. These retain the slide-valve, as a sliding fit between them, and in order to keep these nuts in position on the screw, jam-nuts are used at both ends, so that there are four nuts in all, two in front, and two behind. The longitudinal hole through the body of the valve, for the spindle to pass, is made of an oval shape, to allow the valve to adjust itself, so as always to press against its seat. In small models, instead of using nuts, it is easier to make a longitudinal saw-cut along the back of the valve, and file down a piece of the valve-spindle quite flat, leaving a collar at each end. Slip this part into the saw-cut, the collars hold the valve in place, and the saw-cut allows adjustment of the valve to its seat. The reciprocating movement of the slide-valve is derived from the circular motion of the centre of the eccentric-pulley, through the intervention of the eccentric-strap, the eccentric-rod, and the valve-spindle.

The **valve-spindle** is made of iron wire, turned circular, about half the thickness of the piston-rod. One end is attached to the valve, as mentioned above, and the other end is screwed into a block, which by means of a pivot unites with the eccentric-rod; the spindle passes through a steam-tight stuffing-box and gland. In large engines, there is a guide to keep the slide-valve in a straight line; but this is not absolutely necessary in small models, the stuffing-box and gland being sufficient for this purpose.

The **eccentric-rod** is an iron wire, circular or flattened, screwed at one end into the eccentric-strap, and at the other end having a hole passing through it for the spindle-pin; instead of this, however, it may be forked at this end.

The **eccentric-strap** is a brass casting, either cast

in two halves, or if solid, it must be equally halved by a saw-cut, going right through the centre of the hole or aperture cast in it, entering and coming out in the middle of the lugs, through which the bolts pass. It must be filed smooth all over, and, with a round file, the inside of each half filed out, until it

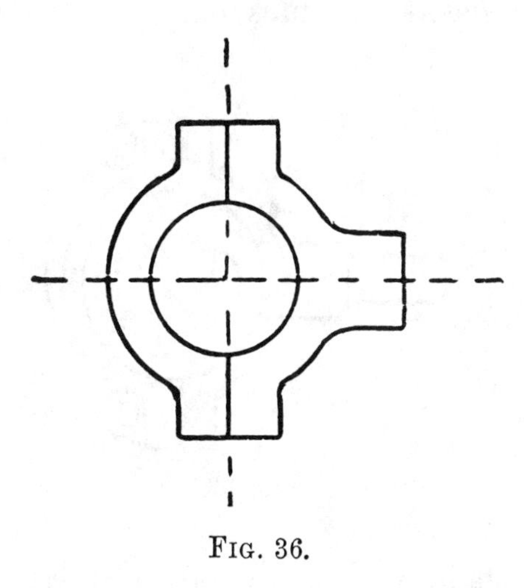

FIG. 36.

accurately fits upon the eccentric-pulley, when the two halves are brought together, and yet allow the eccentric to revolve freely inside. After fitting, drill a hole through each lug and pass two bolts through, which are tightened up with nuts; a small hole can be drilled, and an oil-cup screwed in for oiling the eccentric-pulley (see Figs. 36 and 37).

The **eccentric-pulley** is a circular brass casting, or is turned out of a bit of brass rod: the former is adopted for the larger sizes. This has a groove cast around the circumference, as well as a tenon or projecting boss attached to one side, for the purpose of gripping it in the lathe-dog when chucking; the other side has, sometimes only, a second boss cast

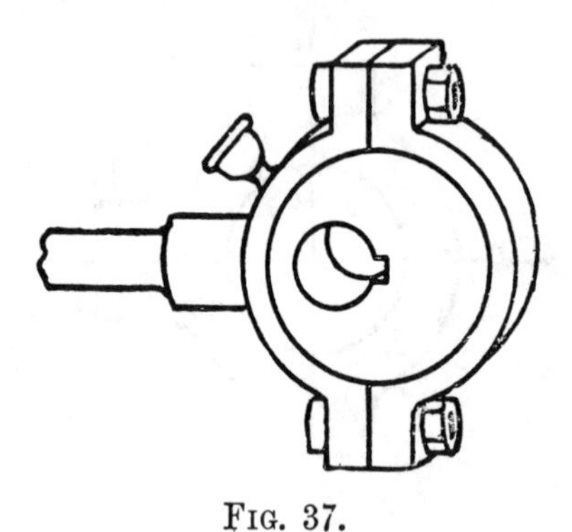

FIG. 37.

upon it, made to correspond with the centre of eccentricity, and is for the purpose of enabling the eccentric to be fixed to the crank-shaft by means of a set-screw, passing through it at right angles to the shaft. If there is no boss, the eccentric must be fixed to the crank-shaft by means of a key, and a key-way must be cut out with a small triangular file. In real engines eccentrics are keyed to the shaft. To turn the eccentric, chuck it in the lathe, smooth the flanges, clean out the groove, and face up the sides. After turning, saw off the chucking tenon, and file the edges of the saw-cut quite flat, lay the eccentric on a flat table or board (if there is no boss) with that side up, having the true centre marked on it (the other centre is lost by cutting off the tenon); measure, from this centre, a distance equal to half the travel of the slide-valve across the port face, mark the place with a pair of compasses (having their legs fixed by a set-screw), punch a hole over this mark—this is the centre of eccentricity, and a hole must be drilled right through the eccentric, which shall remain quite parallel with the centre of the true axis of the eccentric, or in other words, with that of the crank-shaft. If there is a boss, the pattern should have been so carefully made that the centre of this boss will exactly coincide with the centre of eccentricity, then this centre is easily found. To bore the eccentric, for the axle to pass through, is rather difficult, so as to keep the hole quite true; but can be readily done, if the amateur possesses an upright vertical bench drill, as all he has got to do is to lay the eccentric flat on the bed, keep it so, choose the proper-sized drill, and bore out there and then. If only a hand-drill is his, then the eccentric must be fixed in the vice, and bored as straight through as possible; a small round file will help to bring the hole back "true," if it should be a little off, *i.e.* not quite parallel with the true axis of the eccentric; if it was left so, the eccentric when put on the crank-shaft would wobble, and catch against the eccentric-strap: it must run dead true when on the shaft. Th hole drilled must be just so large, as to allow the eccentric to slide along the crank-shaft. If there be a boss, it can be turned by mounting the eccentric on a temporary shaft in the lathe, or it may be smoothed over with a file. The key-way should always be cut in the eccentric opposite the part where the metal is thickest. The throw of an eccentric is the distance between the centre of the crank-shaft and the centre of the eccentric-pulley. The latter form of pulley (for small engines) is turned either singly, or in pairs, out of brass rod, and sawn apart. An eccentric, $\frac{11}{16}''$ diameter over all and $\frac{1}{4}''$ wide, should have a groove $\frac{1}{16}''$ deep and $\frac{1}{8}''$ wide, with $\frac{1}{16}''$ collars on each side. In order to insure the hole being quite true (for the crank-shaft) that passes through, find the centre of eccentricity on one side and punch the hole; place the eccentric in the lathe, with the dead-centre point entering the punched hole, bring up the live-centre point so as just to touch but not mark the eccentric, turn the lathe slowly round, and the eccentric will tend to revolve between the centres; if it does so, keeping at right angles to the central axis of the lathe-centres, and does not wobble, it is right. Then screw up the live-centre, so as to make a mark at this part, remove and punch a hole on this mark, fix the eccentric in the vice, drill by hand, or on the lathe, through half-way from each side, meeting at the centre. The hole thus drilled must be true, if properly done; if it should be a little way off, though only fully $\frac{1}{8}''$ diameter, it may be made true, by passing through it a small round jewellery file, such as is used by watchmakers, and filing out where required. The eccentric is attached to the crank-shaft by means of solder, or a boss for a set-screw can be made and soldered on upon one side.

In addition to the eccentric-strap mentioned above, we will describe another kind which can be adopted for small engines, especially locomotives, with very good results. To make it, take a piece of thin sheet brass, cut a long narrow strip from it, much longer than the circumference of the eccentric-pulley, and of a width that will be an easy fit in the groove of the pulley. A short distance from one end, bend the brass at right angles, so as to form a lug, then mould the rest circular, to the same diameter as the pulley, for rather less than one-quarter of its circumference, and with a pair of pliers make a small loop on it (see Fig. 38), and from this loop continue moulding the circle to the same diameter as before. When the circle is completed, bend the remainder of the strip at right angles, and cut off short, to form a lug. In this way, there are two lugs formed at the top of the strap which, when in position on the pulley, lie against each other, and through these the bolt passes for holding the strap on the pulley. Drill a hole through the middle of the loop, take the eccentric-rod, screw two nuts on to one end, remove these, solder one nut inside the loop, screw the other nut on to the eccentric-rod, leaving an end, which must be passed through the loop, and firmly screwed into the nut soldered inside it. Screw the other nut upon the rod up firmly against the loop, and these two nuts jam each other, and hold the rod quite tight in its place on the eccentric-strap; a little solder will also help. Or (instead of nuts) the rod can be put through the loop and soldered with hard solder. This form of strap is much lighter than a casting, and as the loop prevents the strap from rubbing at every part upon the eccentric-pulley, it works with very little friction, and does well for small locomotives, where saving of weight and friction is a great matter. The strap has a good deal of spring in it, and when the bolt is removed slips very readily off the eccentric-pulley. This strap must be made quite circular and a good fit on the pulley, for, with a loop at one end, there is apt to be some play antero-posteriorly on the pulley, which it must not have. To fit it on properly, turn a piece of hard wood to the same diameter as the eccentric-pulley at the bottom of the groove, tighten up the strap with the bolt, and pass the strap over the end of the piece of wood. After removal from the lathe, with pliers tighten the loop, till the strap becomes an easy fit on the wood, and quite circular.

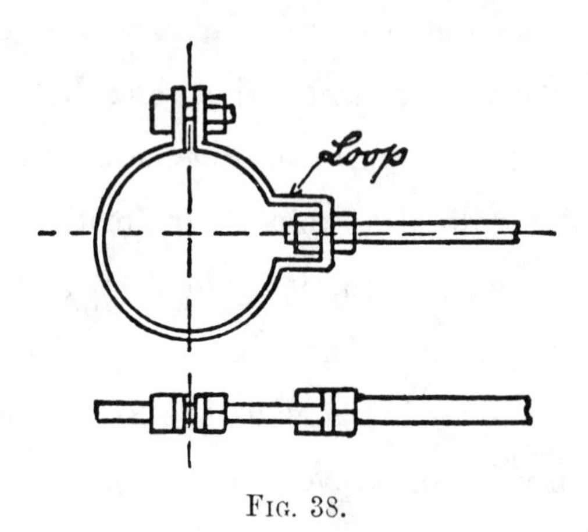

FIG. 38.

Remove the strap, take out the bolt, slip it over the pulley, and tighten up the bolt, with the nut, when the strap will be found an exact fit on the pulley, and its action on the slide-valve perfect. It can be filed slightly all round on the inside, with a round file, as this makes it work more easily (see Fig. 38).

Oil-cup. This is a casting, or else is turned and bored out of a piece of brass rod, and screwed into its place (in a hole drilled to receive it) on the guide-bars, bearings, connecting-rod ends, etc. Those for the cylinder, with single or double cocks, had

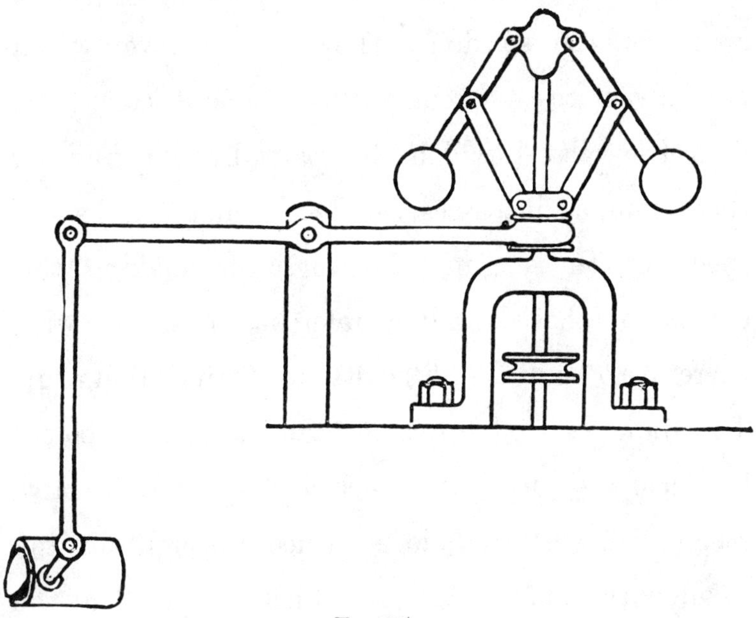

Fig. 39.

best be purchased ready made, as the accurate fitting of cocks is troublesome.

The **governor** (see Fig. 39) **is a device for regulating the speed of the engine.** Different kinds are used; perhaps the easiest to make is the "throttling-governor." It consists of two balls suspended from the top of a vertical revolving spindle by jointed rods, which revolve with it. When the engine runs, they revolve at a certain distance from each other, and if the speed increases, the centrifugal force causes the balls to fly further apart, and, in so doing, they raise a lever which partially closes a circular disc of metal called the "throttle-valve," which, by partially shutting off the steam, as it passes to the cylinder, checks the speed of the engine. Castings for making this can be purchased, and by referring to the drawing there can be little difficulty in fitting them up. The frame is filed up smooth and flat; it is bolted by means of feet to the bed-plate or slide-valve casing. A vertical hole is drilled through the top to allow the spindle to pass, which is tapered at the foot, and rests in a small hole drilled in the bed-plate, so as to keep it vertical. Sometimes a small bevel tooth-wheel is fixed on this spindle, between the legs of the frame, and gears with another tooth-wheel, fixed on a spindle at right angles, which passes through a bearing in one leg of the frame, and carries a small pulley outside. An easier plan than this is to turn a small pulley, and fasten it by soldering or otherwise to the vertical spindle; this pulley is horizontal, and is driven by a band from the crank-shaft. The opposite end of the spindle carries a brass cap, firmly screwed to it, having a projecting lug on either side. We must now turn the balls and their spindles on the lathe, and slot a piece out of the free end of each spindle, and pivot a ball to each lug. The sleeve is simply a piece of brass, which can slide up and down on the spindle; this carries two small lugs at the top, one on each side, and underneath these is a kind of pulley with a broad groove forming a collar (this sleeve must be turned on the lathe); two forked rods or links are pivoted to the lugs of the sleeve at one end, and, by means of the forks, embrace the governor arms about the middle of their length, and are pivoted to them. A long brass rod must be pivoted to a bearing at its centre, so that the forked end engages with the sleeve-pulley, and is raised and lowered by it; the other end is pivoted to a lever connected with the crank of the throttle-valve.

The **throttle-valve** is a brass disc, turned circular, so as to fit its case, left thickest at the centre, in order to be drilled, and fixed upon a central horizontal spindle, by means of a small set-screw. The inner end of this spindle is supported in a round hole drilled in the pipe, whilst the outer end projects through a steam-tight stuffing-box and gland, soldered into the opposite side of the pipe. Outside, a small crank is attached to this end, which is worked by rods from the governor-sleeve. When the throttle-valve is closed, as when the balls hang nearly vertically downwards, the valve bears all round on the inside of a short piece of pipe or casing, which has been bored out truly cylindrical. This casing must be screwed or soldered between the steam-cock or stop-valve and the slide-valve chest. When the governor has been bolted down in position, and a small pulley been rigidly fixed on the crank-shaft, a piece of cord or string is passed from this pulley to the governor-pulley and tightened up; then when the crank-axle revolves, the governor will revolve, and cause the balls to fly apart, and open the throttle-valve.

The governor is rather a complicated piece of mechanism to construct, and may be omitted in engines of less than $1\frac{3}{4}''$ bore cylinders. For this size, the balls may be of iron, $\frac{3}{4}''$ diameter, and arms $3''$ long, of $\frac{5}{16}''$ iron, with a coiled spring on the spindle to push down the sleeve. The spindle may be $\frac{7}{16}''$ to $\frac{1}{2}''$ thick; the spring must be chosen according to the speed adopted; the pipe that holds the throttle-valve may be made $\frac{7}{16}''$ to $\frac{1}{2}''$ diameter inside.

CHAPTER IV.

FITTING UP THE ENGINE, AND SETTING THE SLIDE-VALVE.

Now that all the different parts of the engine are finished, we must fit them together, and in order to show how this is done, we will suppose we are going to make a horizontal engine.

The cylinder stuffing-box must first be packed, to render it steam-tight, and to do this the gland should be unscrewed, and a small piece of tow, soaked in oil, wound twice or thrice round the piston-rod between the stuffing-box and the gland, and pushed into the former with a wire, and the latter screwed firmly up on the packing to compress it, and given a turn backwards to make the piston go more easily. Work the piston backwards and forwards by hand, using plenty of oil, and it will soon work quite easily; pack the slide-valve stuffing-box in exactly the same way. Take the bed-plate, which, if it has been purchased from the model-maker's, will have the positions of the various parts marked upon it; but if not, you must set out the positions of the cylinder, guide-bars, plummer-blocks, etc., from working drawings, which in the case of small models could be drawn to full size. The cylinder feet or lugs must be filed flat, and drilled for the bolt-holes; the position of these holes must be accurately marked on the bed-plate, and drilled for screwing in the bolts, and the holes in the cylinder feet should be widened with a rimer, so as to pass easily over these bolts, and then the cylinder is fixed by nuts to the bed-plate. Be sure that the cylinder is filed perfectly flat on the bed-plate, and that it is so placed, that the piston-rod is perfectly parallel with the bed-plate when fully pulled out. Holes must be bored in the ends of the guide-bars, and corresponding ones made in the bed-plate, at the proper distance from the cylinder corresponding with the stroke, the two sets of bars having the piston-rod truly in the centre between them; bolt these firmly down to the bed-plate, the nuts being underneath. See that the cross-head and its "slippers" slides easily along between the top and bottom bars. The crank-shaft should be placed in the bearings, and these laid on the bed-plate. The shaft must be set at right angles to the piston-rod: mark the position of the bolt-holes, which pass through the soles of the plummer-blocks, upon the bed-plate, drill these and bolt the plummer-blocks firmly down; attach the connecting-rod to the crank-pin, by taking out the bolts from the "big end," removing one of the brasses, and slip it over the crank-pin; put on the "step," and tighten up the bolts. Remove the back cover from the cylinder, push the piston back till it just touches the edge of the steam-port, turn the crank round, until it is nearest to and fair in line with the cylinder, or upon the "inner dead centre," mark on the loose end of the connecting-rod the position of the cross-head pin, remove the connecting-rod from the crank, and drill a hole corresponding with this mark, to be an easy fit on the pin. As mentioned before, the cross-head pin passes through the cross-head, the connecting-rod end, and enters the centre of each of the "slippers." Place the connecting-rod on the crank-pin again, remove the upper guide-bars, take the pin out of the cross-head and the guides, place the free end of the connecting-rod in the slot of the cross-head, push the pin through the connecting-rod and cross-head, place the guide-blocks upon each end of this pin, and bolt down the top guides; slip the eccentric-pulley with its strap and rod over the end of the crank-axle, key the fly-wheel upon its seat, on the crank-shaft near one end, and a small pulley upon the other; screw a steam-pipe with a stop-valve, and an exhaust steam-pipe, into their respective openings, screw the four bolts for holding the slide-valve casing, into the cylinder face, and slip the valve-casing (having packed the stuffing-box and placed the valve on the spindle) over the bolts, after smearing the joint with red lead or inserting a piece of india-rubber, to make a steam-tight joint; fasten this firmly on with two nuts, one at either corner; keep the valve-cover off until the slide-valve is properly adjusted. When measuring to get the proper length of the eccentric-rod, etc., always keep the slide-valve well pressed up against its seat by the finger, or by a temporary strip of brass inserted for this purpose. We must now set the slide-valve, and it is very important to adjust it correctly, or the engine will not work. By "setting" the slide-valve, we mean fixing the eccentric-pulley in a definite position in regard to the crank, that the relative positions of the slide-valve and piston may produce the result of permitting steam to enter and leave the cylinder at the proper times; and in order to do this, the eccentric must be fixed ahead of the crank by a definite angle. We must first get the length of the eccentric-rod, which is done by placing the slide-valve in the middle of the cylinder face, so that the steam-ports are equally covered (the "lap" being equally divided). Now fix the eccentric temporarily, in any position on the crank-shaft, and turn the shaft round until the wide side of the eccentric is in its highest or lowest position, and measure from the centre of the pin in the valve-spindle cross-head to the bottom of the groove on the eccentric-pulley, which is the length of the rod, including the strap. The rod may be allowed a little longer than the actual measurement at first, as it can always be shortened; but its length can only be extended within very narrow limits. The free end of the rod must be drilled, where marked, to take the cross-head pin, and when attached, it should be adjusted to its true length, by screwing it in and out of the valve cross-head. Having fixed the eccentric-rod, turn the shaft round until the "throw" or wide part of the eccentric-pulley is in its furthest position from the cylinder, look how much the steam-port for the back of the cylinder is opened, then turn the shaft again, until the "throw" or wide side of the eccentric-pulley is in its nearest position towards the cylinder, and notice the amount of opening the front port of the cylinder has. If this port be opened wider than the back port was the rod is too long, and must be shortened half the amount that one port is opened wider than the other; if the back port is too much opened, the rod requires lengthening half the difference, by unscrewing it further out of the eccentric-strap. When both ports are equally uncovered, the length of the eccentric-rod is correct. We must fix the eccentric-pulley upon the shaft so that steam shall be admitted to the cylinder at proper intervals during the piston stroke. To do this, bring the piston round to the back of the cylinder, the crank being nearest it on the inner dead centre, loosen the eccentric, and turn it round upon the shaft until the back port is just open and no more, and might admit a very thin strip of paper; then key the eccentric temporarily, and turn the shaft round till the crank is on the outer dead centre and the piston up at the front port: if the edge of the paper just enters the opening in this port, the valve is correctly set. This amount of opening before the piston commences its stroke is called "lead," and if it does not affect the ports equally, the length of the eccentric-rod must be re-adjusted. When the position for fixing the eccentric on the shaft has been found, mark it with a file; this forms the key-seat on the shaft, and must correspond to the key-way in the eccentric-pulley; remove the key that was holding the eccentric temporarily to the shaft, push the pulley over to one side, and file a flat on the shaft. When this is done, put the eccentric back in position, drive the key in firmly, and the whole is complete.

It will be apparent that the centre line of the eccentric must be in advance of the centre line of the crank by a little more than 90°; this "little more" is called the "angle of advance." If there was neither lap nor lead, then the centre line of the eccentric would be at right angles to the centre line of the crank, or the eccentric be only 90° ahead of the crank.

Having set the slide-valve, bolt the back cover on to the cylinder, making the joint tight with rubber or red lead, and do the same with the slide-valve cover. Mount the engine, and screw it firmly down upon a solid wooden block, of the same length and width as the bed-plate, having made holes in the top of the block, into which any nuts projecting underneath the bed-plate may enter; or instead of this, the bed-plate may be fixed upon turned brass pillars, and these driven firmly into a wooden sole, which can be stained or varnished afterwards. After finishing an engine, it is a first-rate plan to connect it temporarily by means of coupling-screws and brass tubing to a horizontal boiler set over the kitchen fire (see page 36), and after steam is up, giving the engine a run of about an hour's duration. In this way, the engine gets into proper working order, and will run smoothly ever after.

CHAPTER V.

REVERSING-GEAR.

Reversing-gear. By this is meant a contrivance fitted on the engine to enable it to run either backwards or forwards. This is carried out in different ways: (1) By a loose eccentric on the crank-shaft (*a*) fixed by a set-screw, (*b*) working against a stop; (2) by a link motion (*a*) having one eccentric fixed on the crank-shaft, (*b*) having two fixed eccentrics on the crank-shaft.

1. (*a*) **A loose eccentric fixed by a set-screw.** We must first set the eccentric on the crank-shaft for running in one direction, according to the rules already given, mark the position of the set-screw on the axle, after the eccentric is properly set, and drill a hole in the axle at this point, into which

the set-screw enters for a short distance, and is tightened up. Then set the eccentric for running in the opposite direction, and drill a hole in the proper place on the axle for the set-screw as before. By simply changing the position of the eccentric, the engine will run in either direction.

(*b*) **The eccentric working against a stop.** The eccentric, being loose on the crank-shaft, is driven backwards and forwards by a projecting feather, soldered or welded to the crank-shaft. A brass ring is fixed on the axle, on the opposite side of the eccentric, against which this latter bears, and is always kept up against the feather. Turning the fly-wheel in either direction causes the fixed stop or feather to come into firm contact with the forward or backward projection on the eccentric pulley; this projection or counter-weight should be firmly screwed or soldered to one side of the eccentric-pulley, and set in the proper position. With a little trouble the eccentric may be accurately fitted, by filing away part of the feather and counter-weight till properly set, and the counter-weight, being fixed to the eccentric-pulley, is carried round by the feather either forwards or backwards, according to the direction in which the crank-shaft forces it. Both of these methods reverse the lead

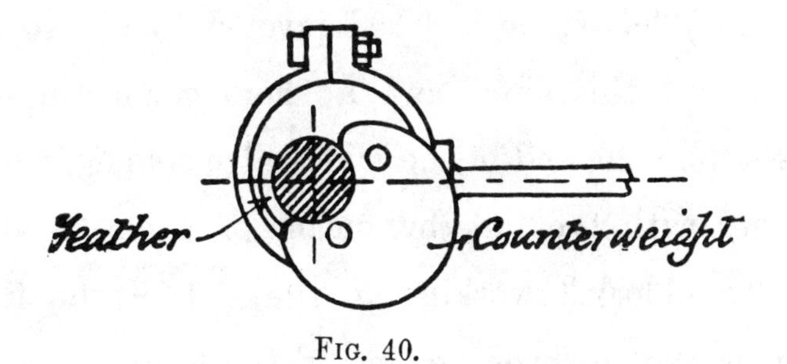

Fig. 40.

of the slide-valve as well as the engine, and act fairly well (see Figs. 40 and 41).

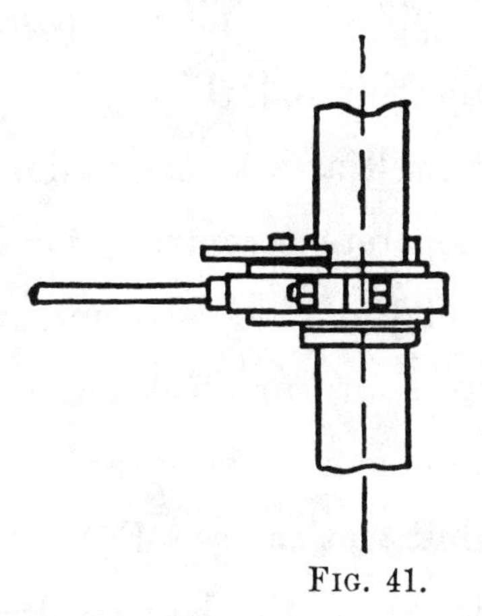

Fig. 41.

2. (*a*) **Link motion with one fixed eccentric.** This is an ordinary curved slot-link, pivoted (at its centre) to a support fixed on the bed-plate, having

the *concave* side turned towards the slide-valve; this link remains fixed. The eccentric-pulley, being fixed on the crank-shaft (for the engine to run in one direction), has its forked rod pivoted to the top or bottom of this link, and a short connecting-rod passes from the end of the slide-valve spindle, which engages with the link by means of a block called the "die-block," working in the slot of the link. By means of a lever moving it, this connecting-rod and block can be slid up and down the link, and though the eccentric-pulley is fixed to run in one direction only, by simply changing the position of the valve-spindle connecting-rod, the engine will run either way, but not so well as with the ordinary link motion (having two separate eccentrics), for it does not reverse the lead of the slide-valve in reversing the engine, and so gives an imperfect action to the valve when the engine is moving backwards.

(*b*) **The ordinary link motion** (see Fig. 42), with two separate eccentrics, gives the best results of all, and we will describe it fully. The eccentric-pulleys are so placed on the crank-shaft, that when one is in the right position for forward motion, the other is in position for backward motion. The eccentrics are connected by separate straps and eccentric-rods to the ends of a link, whose *concave* side is turned towards the axle. In this link a die-block, connected to the slide-valve spindle, is fixed, and can be slid from end to end. By raising or lowering this link

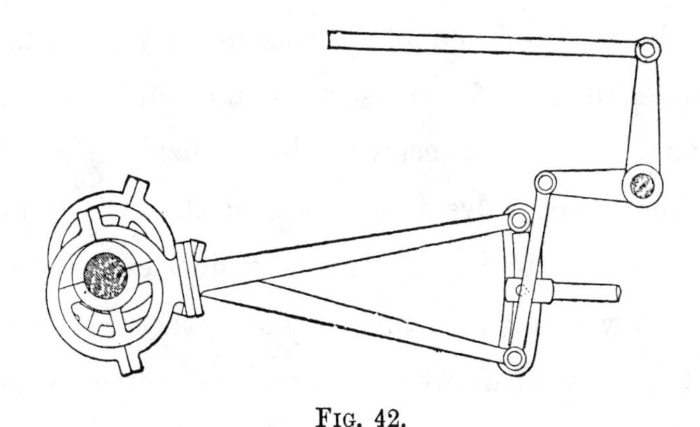

FIG. 42.

(the die-block remaining fixed), so as to bring the valve-spindle in line with one or other eccentric-rod, the motion proper to that rod will be communicated to the slide-valve, and the engine will move backwards or forwards accordingly. The reversing-lever is held in position by a bolt working at its side, and is kept in the notches of a quadrant by a spring, and by means of a weigh-bar, shaft, and levers moves the link up or down as required. If the link be placed so that the block is at the middle of the link, then the engine stops, since the valve is thrown into its mid-position. In real engines, the nearer the die-block is moved towards the centre of the link the shorter becomes the stroke of the slide-valve, and this produces an earlier cut-off of steam and increased grade of expansion, and advantage is taken of this fact to save steam and lessen the amount of coal consumed, by utilizing the full motion of the valve when starting the engine, and then soon notching up the reversing-lever for expansive working. In models where the opening to lead is very small, we cannot take advantage of this plan—in fact it does not work well—but simply design the motion for full gears. This form of link motion reverses the lead of the slide-valve as well as the engine, and so causes the engine to run equally well in either direction. For further drawings, see Sheet No. 11, C, and Sheet No. 8, C.

Details of Valve-Gear.

It is best to make two eccentric-pulleys, and key both separately to the crank-shaft in their respective positions, instead of having them cast together, and one key holding both. Two eccentric-straps and rods are required, each with a forked end, pivoted to the top and bottom of the slot-link respectively. The fork is a brass casting, slotted out, and screwed to the end of the rod; one of the rods must be cranked, by reason of the relative positions of the eccentrics and the link, which cannot all be in the same plane. The slot-link requires great care in making, to get it accurate. It may be constructed of sheet brass or iron plate. Take a triangular piece of metal, long enough to include the link, and the centre from which it is struck, punch and drill a hole through; this represents the centre of the axle, and forms the starting-point for setting out the link. The curved centre line of the link is struck to a radius, which equals the distance from the centre of the axle to the centre of the valve-spindle cross-head pin, the slide-valve being set at "mid position," and kept well pressed up against the port faces. With compasses, mark upon both sides of the metal the centre line of the link and all the other outlines correctly, and with a straight-edge draw the radii correctly (see

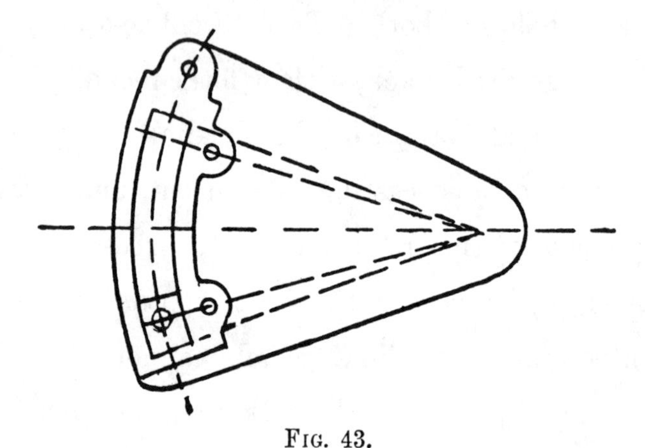

FIG. 43.

Fig. 43). Leave three snugs, two in front near the ends, and one projecting at one end for attaching the lifting-link to; drill and file out the slot correctly, but don't make it too long; mark and drill the hole for the eccentric-rod forks, through the snugs, in such a position that when the die-block is pushed to either end of this slot, a straight line, prolonged from the centre to the central hole in the die-block, will pass through the centre of the holes for the eccentric-forks in each snug. These holes must be drilled in the line of curvature from the centre; and the hole for the lifting-link pin, passing through the end snug, must be drilled in the centre line of curvature of the slot, at some little distance from it (see Fig. 43), before cutting

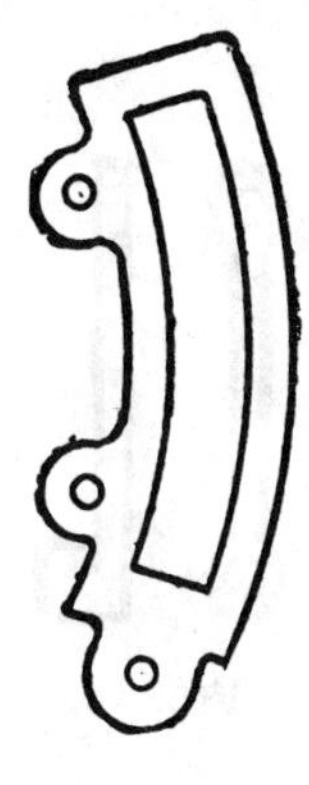

Fig. 44.

the link to outline. From the centre, with a pair of compasses and a straight-edge test the accuracy of these different curvatures and radii upon both sides of the metal; if not correct, they must be filed away until they are. When this is done, proceed to cut the link to outline by drilling, cutting, and filing; smooth it all over with a file (see Fig. 44), then, if carefully made, the link will be found to act perfectly upon the valve. The eccentric-rods are attached to the link by bolts passing through the forks and snugs. The end of the slide-valve spindle must carry a brass fork, which embraces the link but allows it to slide up and down upon its faces; it also embraces the die-blocks. This block (see Fig. 43) is a piece

FIG. 45.

of brass accurately fitted, to slide along the slot-link, convex behind, concave in front; the upper and under borders fit against the top and bottom edges of the slot when in contact with either end of the link. It is made slightly thicker than the link, but of an easy fit, so as to pass between the valve-spindle forks, and a transverse hole is drilled through its centre. Connect the link and valve-spindle together, by placing the die-block in the slot of the link, and pass both between the eccentric-spindle forks, put a pin through both legs of the fork and die-block: the link oscillates upon this pin.

Having now connected together the eccentrics and the slide-valve, let us turn to the lifting-links, weigh-bar, reversing-lever, etc., which provide for the reversing of the engine.

The **lifting-links** (see Fig. 45) are straight or curved, made of soft iron filed flat on opposite sides. In large engines there are two, one is placed on each side of the slot-link; in models we may use one only. One end of this link is pivoted to the top, the bottom, or the centre of the slot-link, the other end to the weigh-bar lever. The end that engages the link should have a projecting boss upon the inside, to keep it at a fixed distance from the slot-link, so as not to interfere or strike against the slide-valve fork and eccentric-forks when lifting up and down. The same thing may be accomplished by placing washers between them.

The **weigh-bar** is a piece of iron wire, fixed in bearings, placed at a suitable distance from the slot-link, so as to allow of the latter moving up and down freely. It has firmly fixed to it a lever which at one end pivots to the lifting-link, and at its other end is prolonged behind the weigh-bar for a certain distance, and carries a balance weight—consisting of a piece of brass, cored out for the lever, which passes through it—which is fixed in position by a set-screw. There is another short lever keyed to this weigh-bar, which by its free end is pivoted to a rod, which again is pivoted to the reversing-lever working in a quadrant. This quadrant is a small casting, having feet or lugs, by which it is bolted to the bed-plate, or the foot-plate of a locomotive. The quadrant must be correctly curved at the top to suit the locking-bar, and notches cut in it to correspond with full-forward, backward, and mid-gear. (The positions for these notches should be found out by trial with steam afterwards.)

The **reversing-lever** is an iron rod forged or filed out of iron wire, with a turned handle on the top. It is pivoted to the foot of the quadrant, and a rod is pivoted to the lever (see drawing) a short way under the "notch-plate," which by its other end engages with the short lever on the weigh-bar.

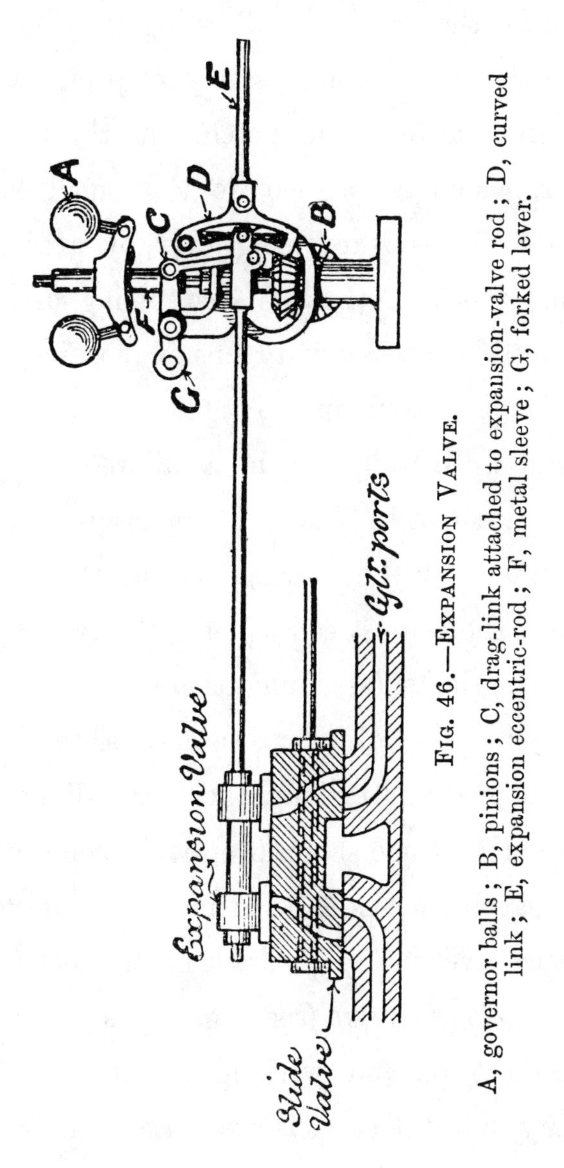

FIG. 46.—EXPANSION VALVE.

A, governor balls; B, pinions; C, drag-link attached to expansion-valve rod; D, curved link; E, expansion eccentric-rod; F, metal sleeve; G, forked lever.

A small spring attached to the handle of the

reversing-lever, causes the locking-bar to be thrust down into the notches. Pressing a small lever towards the handle compresses the spring, and lifts the locking-bar out of its notch. At the lower end the bar, which is pivoted to the small lever, is held close to the reversing-lever by a ring that surrounds both of them. Good fitting of all pins and joints is necessary to ensure accuracy, for if the parts be fitted up too tight, or too slack, the gear will work badly and be a failure.

In a locomotive there is precisely the same arrangement of the gearing, but as there are two cylinders there must consequently be two slot-links, four lifting-links (assuming there is one on each side of the slot-link), four eccentrics, two counter-balanced levers attaching the lifting-links to the weigh-bar, and one short lever again connecting the weigh-bar, by means of a rod pivoted to it, with the reversing-lever in the quadrant on the foot-plate.

An **expansion-valve** (see Fig. 46) is a valve fitted steam-tight to, and working on, the back of an ordinary slide-valve. It is used in engines for obtaining high grades of expansion of steam, which the ordinary slide-valve is not well fitted to give, as sufficient opening to steam cannot be obtained with it for an earlier cut-off than, say, half-stroke, without unduly increasing the travel and the lead. In marine engines the expansion-valve is worked by a separate eccentric fixed upon the crank-shaft, the eccentric-rod of which connects to a straight link moved by a horizontal screw, and a pair of bevel-pinions, so that different grades of travel—and consequently of expansion—may be obtained. The nearer the eccentric-rod is placed in line with the expansion-valve spindle, the greater the travel and the later the cut-off. But the plan we are going to adopt here is to connect the expansion-valve spindle direct with a Hartnell's governor, as is done upon large engines by Messrs. Marshall and Co. of Gainsborough, and probably by other makers as well. A model expansion-valve, though not required to regulate the steam, if well fitted has a fine appearance, and makes the engine look more finished.

The **slide-valve** (see Fig. 46) somewhat resembles the ordinary slide-valve, by being box-shaped and attached to the spindle in the same way, but is much longer, and during its travel it never leaves the ports uncovered at the ends, for steam to pass into them round its edges. It is so constructed that all the steam passing through the cylinder-ports must first pass through the valve itself, and it contains for this purpose a port near either end. Make a brass slide-valve as before, and, in order to keep clear of the spindle, drill near each end two holes, one on each side of the spindle in a slanting direction, so that they join underneath the spindle, and come out on the face as one hole, near each end. These ports are for the admission of steam to the cylinder; the valve exhausts in the same way as before. This valve-spindle passes through a stuffing-box, and connects, by the eccentric-rod, with an eccentric-pulley keyed to the crank-shaft, and set according to the rules already given, so as to admit steam to the cylinder when required. This valve regulates the point of admission, release, and compression, but not the cut-off; this last is regulated by the small valve that works on the back or top of the slide-valve, and which is directly connected with the governor. It consists of a single brass plate with two faces (the best arrangement in a model), or two plates working steam-tight on the back of the slide-valve. These plates are kept at a fixed distance apart by collars fixed on a spindle, which passes through a second stuffing-box in the valve-casing. The action of these plates is to cut off the steam at any desired part of the piston's stroke, and prevent it from passing through the slide-valve and cylinder-ports into the cylinder. The free end of the expansion-valve spindle has a die-block, which engages with a curved link attached to the governor.

A Hartnell's governor has some resemblance to that on page 119. It is driven by a pulley keyed on the crank-shaft, and this pulley is connected by a belt or cord with another pulley keyed on a vertical spindle, which by its other end supports the balls, which are fixed to bell-crank levers; these balls revolve with the vertical spindle. The inner ends of the two bell-crank levers bear on a spiral spring contained in or above the metal sleeve. On the lower end of this sleeve is fixed a double collar, engaged by a forked lever, suspended from which is a drag-link, the lower end of which is attached to the end of the expansion-valve rod; the end of this rod engages a die-block, which can be drawn from one end to the other of the link-slot, which is curved towards it, and to the centre of this link is attached the expansion eccentric-rod, the other end of which is attached to an eccentric-pulley fixed on the crank-shaft.

Action.—When the speed of the engine exceeds the normal, the balls fly outward and compress the spiral spring, lifting the brass sleeve, the drag-link, and the expansion valve-rod with die-block, towards the upper end of the curved link, thus diminishing the travel of the expansion-valve, and cutting off steam much earlier, which reduces the speed to the normal. When the speed falls below the normal the opposite takes place, for the spiral spring overcomes the compressive pressure of the bell-crank levers and presses down the sleeve, the drag-link, and the expansion valve-rod with die-block towards the lower end of the curved link, thus increasing the travel of the valve, and causing a later cut-off, which tends to increase the speed again. This governor is more accurate in its action than the throttling one, and does not seem very difficult to construct, though we think the amateur will require to make patterns for castings to fit up large-sized models of this piece of mechanism, which any one who could fit up the throttling-governor and link-motion would be able to do.

A small-sized governor for a $\frac{1}{2}''$ bore cylinder engine might be constructed without castings. Make the vertical support by sawing and filing a brass block to shape shown, turn the balls and arms out of brass rod, make a wire spindle, and procure small bevel pinions and a spiral spring from any watchmaker. The links can be cut out of strong sheet brass, the curve of the slot-link being struck at a radius from the centre of the slide-valve when set at "mid-position"; the various rods and links can be pivoted on each other with fine copper rivets, and with small bolts and nuts. When finished, the governor is bolted by the sole to the bed-plate of the engine, and the expansion-valve and eccentric-rods can be made of any desired length, according to the distance between the cylinder and the crank-shaft.

Joy's valve-gear (see Fig. 47). This form of gearing does away with eccentrics and the ordinary link motion. It is applied both to marine engines and locomotives, and a working model of it would not be very difficult to construct. The motion of the slide-valve is carried out and the reversing of the engine effected by a number of links and connections between the connecting-rod and the valve-spindle. At a joint C, on the connecting-rod CR, is attached a double link CL; about one-third the distance of this double link is attached a pair of double links VL, their upper ends first being attached to a sliding-block SB, which works in a curved slot-link FF, and then are pivoted to the valve-spindle connecting-rod VSL, which connects with the valve-spindle and the slide-valve. The lower end of the double links CL is connected to a radial rod AL, which is pivoted to a fulcrum B, bracketed to the frames. The slot-link FF (the radius of which is struck from the centre of the slide-valve when placed in "mid-position") is free to move to the right or to the left, in bearings carried on the engine frame. We see that while the point C of the con-

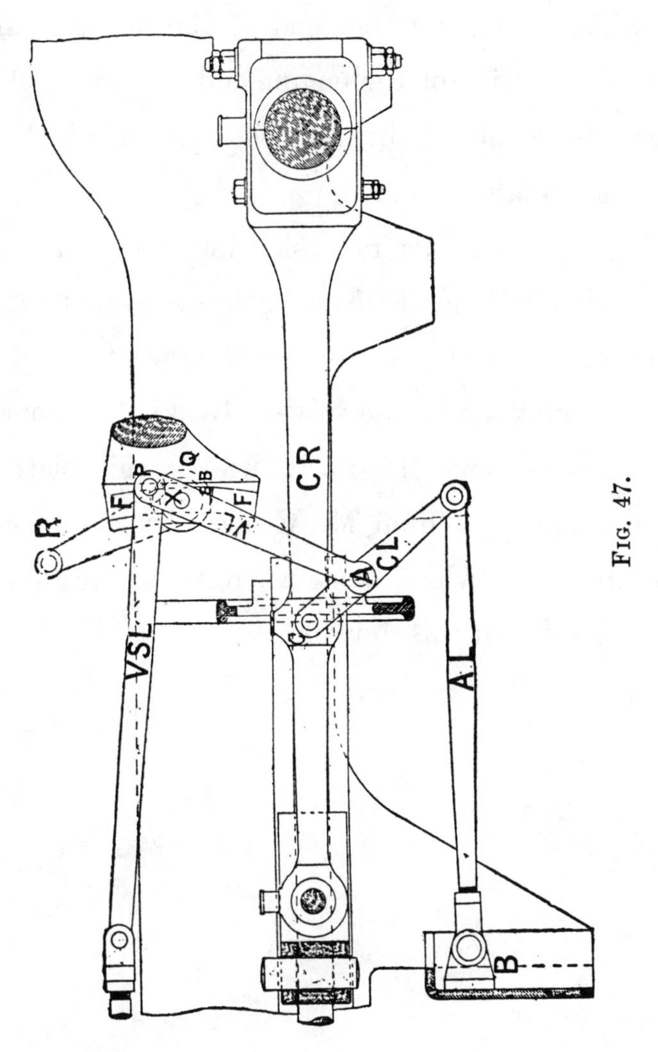

Fig. 47.

necting-rod (to which the connecting-link is attached) describes an oval, the point A of the latter describes a flattened ellipse, thereby imparting an equal motion to the point X. The motion is reversed by the lever R, which is fixed to one end of the quadrant-shaft Q; the direction of the engine and the travel of the valve is regulated by the position in which the quadrant-shaft is placed.

Fig. 48 is a diagram showing the action and expansive working of steam in the cylinder at various parts of the stroke, and the relative positions of the piston, crank, and slide-valve. By kind permission of the publishers, Messrs. Whittaker, this plate has been reproduced from Mr. Cooke's book on *British Locomotives,* to which work we refer our readers for a description of this drawing.

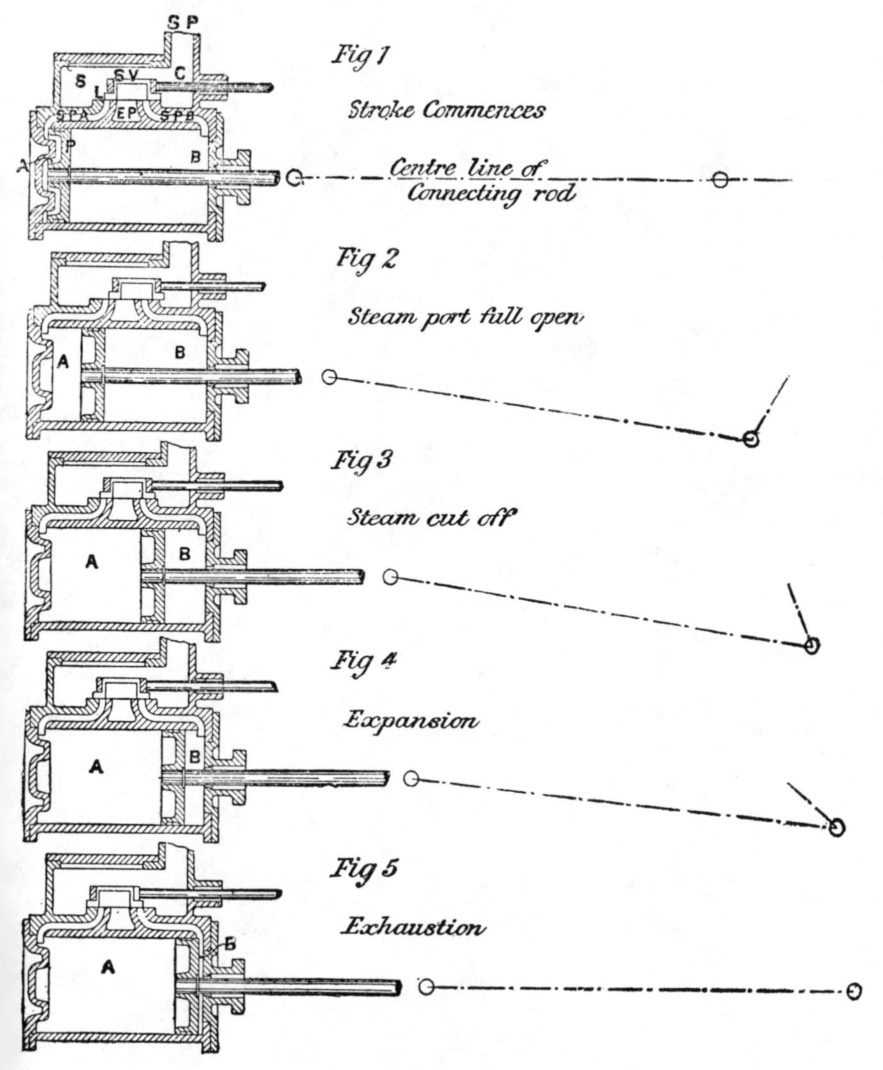

FIG. 48.—Action of Steam in Cylinder, etc.

PART II.

DIFFERENT TYPES OF ENGINES: STATIONARY, LOCOMOTIVE, MARINE.

CHAPTER VI.

BEAM, HORIZONTAL, AND VERTICAL ENGINES.

HAVING described the engine in detail, how to make and fit together the different parts, we will now describe and illustrate a few different types of engines; and any one who can make a horizontal engine, will find little difficulty in constructing any of the forms we take up, as with slight variation they are all fitted up in a similar way.

We give working drawings of a high-pressure **beam engine**—condensing apparatus and pumps omitted—(see Fig. 49 and Sheet No. 3). This is undoubtedly the oldest style of engine in existence, Newcomen having constructed an atmospheric beam engine for pumping purposes as early as the year 1712. This engine, in the hands of Watt, underwent great improvements, and was made double-acting by him in 1784: this style is similar to those constructed at the present day.

By reference to the drawings it will be seen that the cylinder is bolted to the bed-plate so as to

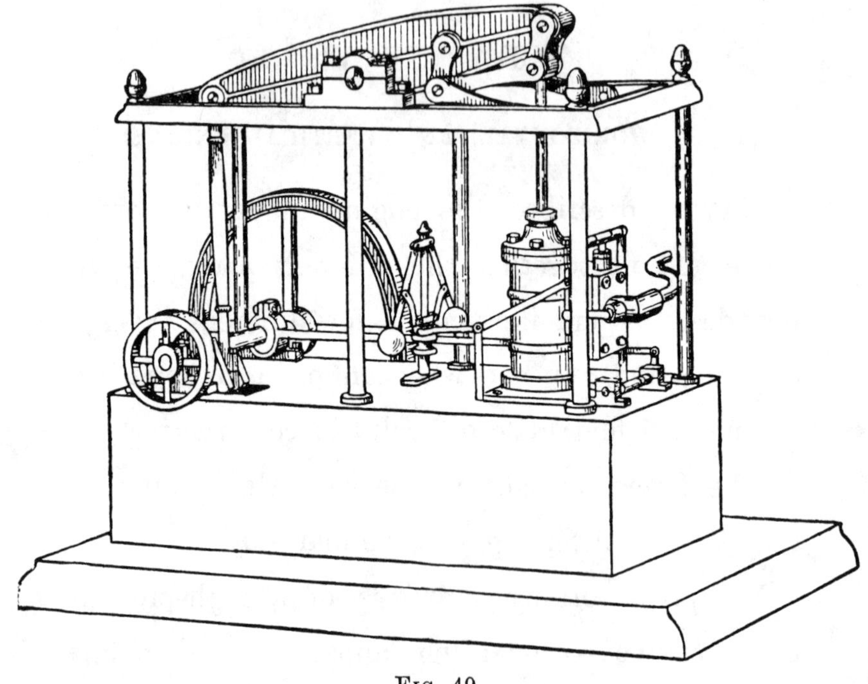

Fig. 49.

stand vertically, and the cross-head is attached to one end of a large beam, which oscillates by its centre in bearings fixed on supporting columns; the reciprocating movement of the piston is communicated to the beam, and by it is transmitted through the connecting-rod at its opposite end to the crank-shaft and fly-wheel. The end of the beam moves in the arc of a circle, while the piston's motion is truly vertical; and to get over this difficulty use is made of the celebrated invention of James Watt,

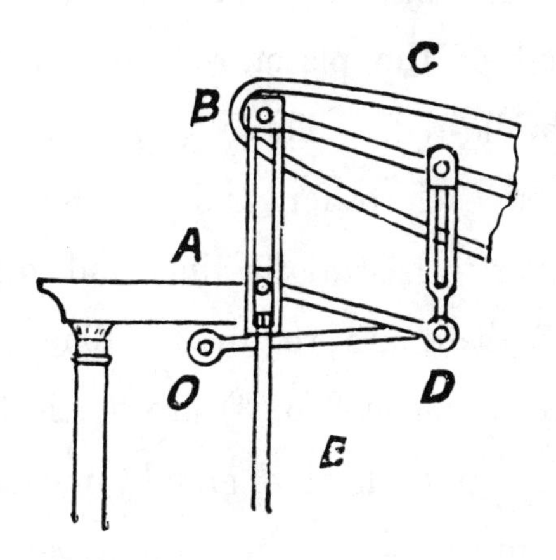

FIG. 50.—The "Parallel Motion."

which is known as the "parallel motion" (see Fig. 50).

From what we have just said, it will be seen that it is impossible to join the cross-head and the beam directly together, so they are connected by means of a short rod AB, which with the other two rods DA, DC, together with part of the beam CB, form a jointed parallelogram, the angles of which vary according to the position of the beam; the angle D is connected by a joint with the end of a small rod DO, movable on a pivot at O. *Effect.*—If the beam be horizontal, and the end B rises, A will pass to the left by the beam's action, and to the right by the action of the small rod DO, which, by checking the movement of the piston-rod to either side, is called the bridle-rod. These two opposite movements balance each other, and so the piston-rod keeps in a vertical straight line, and other than this, no guide-bars are required. The slide-valve and spindle (see Sheet No. 3) are vertical; rods A pass vertically on each side, and by means of bell-cranks B connect with the eccentric-rod, which works at right angles to the slide-valve. A throttling-governor is used to regulate the admission of steam to the cylinder, and this is driven by a pulley on the driving-axle.

Castings for making a beam engine may be purchased from the model-maker's, the names and addresses of a number of whom will be found on page 309. Either a vertical or a locomotive boiler can be used to drive this engine.

The **horizontal engine** (see Sheet No. 4). After what has been said, we will not dwell long upon this type of engine. The working drawings give an elevation, a plan, and an end view of a horizontal centre-crank engine. The cylinder is $\frac{5}{8}''$ bore, $\frac{7}{8}''$ stroke, fly-wheel $3\frac{1}{2}''$ diameter. The engine is fitted with a throttling-governor, double iron motion bars, grease and blow-off cock, and oil-cups. The exhaust-pipe comes out under the bed-plate, and this latter is slotted out underneath the axle to allow of the eccentric-pulley and crank passing round when revolving; the bed-plate is supported by six turned pillars erected on a wooden sole. A drawing of the bearings is given full size. The centre crank-axle can be a built-up one, screwed and soldered together with brass crank-webs (see page 104).

Sheet No. 5 gives an engine of much simpler construction than the above. The cylinder rests upon two feet, which raise it so high, that the connecting-rod when revolving keeps clear of the bed-plate, and no slot is required in the latter. Two circular guide-bars of iron wire are screwed, one at the top and the other at the bottom, to the front cover of the cylinder, their free ends being soldered into a brass ring which is supported upon the bed-plate. The cross-head carries two guide-bars, which are slotted out, to be an easy fit between the bars when working; the connecting-rod is forked at the cross-head end; the crank is placed at one end of the shaft; the eccentric-strap is made of sheet brass, and contains a loop through which the rod is screwed (see page 116); the exhaust-pipe passes vertically upwards, and the steam-pipe carries a coupling-screw for attaching to a boiler.

A horizontal engine is the easiest of all model engines to fit up, and when made has a fine appearance. Castings for making it can be had from any of the model-makers, and vary in price from a set with a ½″ bore cylinder at 1*s.* 6*d.*, up to a small-power engine (2″ bore cylinder) with castings, including governor, pump, and forgings, at about £2. Larger sizes can also be had.

The castings included in a set are a cylinder, two cylinder covers, piston, two stuffing-boxes and glands, guides, slide-valve case, two bearings, eccentric-pulley, eccentric-strap and rod, connecting-rod, cross-head, fly-wheel, pulley, and slide-valve. The larger sizes have in addition governor and force-pump castings; the bed-plate, pillars to support it, and crank-shaft forgings are extras, and not generally included in a set.

The **vertical engine** (see Sheet No. 6) is similar in form to the last, but is arranged vertically instead of horizontally, and consequently occupies less floor space. We shall mention the fitting up of the parts only that vary, in this engine, from the analogous parts in the horizontal engine. In the drawings, there is a front and side elevation shown. File the bed-plate quite flat, drill holes for the cylinder bolts to pass through, corresponding with the holes in the covers and cylinder flanges, bolt down the cylinder with the cover to the plate, or omit the cover and let the bed-plate form the bottom cover for the cylinder.

Be careful that a line passing through the centre of each guide-bar and the centre of the piston-rod, is at right angles with the centre line of the crank-shaft. File the standards smooth and level, and finish as described on page 100; drill the feet of the standards, and make corresponding holes in the bed-plate for the bolts to hold them down by. The iron wire guide-bars are screwed into the front cover of the cylinder, and have the piston-rod exactly in line between them when drawn out; the free ends of the bars are screwed into two brass supports, the ends of which pass through the standards, and are held in position by nuts. The cross-head has the slippers cast upon it, one on each side; these must be filed over, so that when the cross-head is screwed on to the piston-rod, the slippers are a sliding fit between the guide-bars. The connecting-rod is forked, and embraces the cross-head, a pin passing through both; the other end is fitted with brasses as described on page 97, and is then put over the crank-pin; this pin is screwed into the crank-web, and has a small washer on each side to keep the connecting-rod "big end" from rubbing on the crank-webs. For the construction of the crank-shaft, see page 103. When the engine is put together, the bearings must be level with each other, and the shaft run quite easily. Fit up the eccentric, fly-wheel, and pulley, key these in position on the crank-axle, pack the stuffing-boxes, and bolt the bed-plate down to a block of wood, and the engine is ready for steam.

An **inverted vertical engine** is shown upon Sheet No. 7, which contains the working drawings. In this type, the cylinder is raised above the crank-shaft, and is bolted to the guide-blocks. The bed-plate is a casting, either resting on feet or bolted to a block of wood. It is slotted out, to allow the crank with connecting-rod end to pass through, when revolving. The bearings are of brass, bolted by feet to the bed-plate. The guide-blocks, the front cover of cylinder, and stuffing-box are in one casting, and these are bolted to the bed-plate. In fitting up, find the centre of the front cover, punch and drill a hole vertically through for the piston-rod, coming out in the centre of the slide-blocks; turn this casting upside down, and widen the hole from the stuffing-box end for about three-quarters of its depth, fit on a gland (see page 85), screw the piston-rod into the cross-head, after the guide surfaces have been filed flat, file the inside of the guide-blocks smooth, insert the cross-head between the guides, so that the piston-rod passes up through the stuffing-box, the gland, and the front cover. After packing the piston stuffing-box, screw the piston and the follower (if there be one) on to the piston-rod firmly, then take the cylinder and slip it over the piston; see that it rests on the front cover equally all round, and the piston works up and down easily (this part of the fitting is rather difficult, unless the cover has been faced on the lathe). When this end is attained, and the cylinder is correctly set, mark and drill bolt-holes in the cover, to correspond with those already made in the cylinder flange. If the bolts be screwed into the cylinder flange, the holes in the cover must be widened with a rimer so that the cylinder will pass over the piston, and all the bolts (in its flange) pass easily into their respective holes. Next tighten up the cover with nuts. If the bolts be screwed into the cover, then the holes in the cylinder flange must be widened, and nuts put on upon the top of the flange—either plan will hold the cylinder.—After doing this, there will be no difficulty experienced in finishing the engine.

The castings for a vertical engine consist of the

following parts: a cylinder, two covers, slide-valve and case, piston, two glands, cross-head, connecting-rod, guide, two standards, eccentric and strap, crank-shaft, fly-wheel and pulley; and these vary in price from 3*s.* 6*d.* up to £1, according to the size chosen; the bed-plate, columns (if any), piston, and slide-valve rods can be had at an extra cost.

CHAPTER VII.

TRACTION, PORTABLE, AND SEMI-PORTABLE ENGINES.

A traction engine. A perspective view of this engine is given in Fig. 51, as well as working drawings with dimensions (see Sheet No. 8, A, B, and C). The boiler is made in the same way as the locomotive boiler described on page 37. It is made of sheet copper, the barrel and outer fire-box of No. 24 sheet copper, and the internal fire-box, fire-tube, and tube-plate of No. 21 sheet copper, all the joints in it being brazed. The smoke-box is made of tin, circular in shape; it surrounds the front of the boiler for about $\frac{3}{16}''$, and is made a very tight fit round the barrel; it may be attached by solder. The funnel is $\frac{3}{4}''$ wide, to maintain a good draught of air through the fire-box. Eight wicks burn in the fire-box, four in a row, and a space of nearly half-an-inch is left on both sides and in front, between the sides of the outer fire-box (below the water-line) and those of the lamp for admission of air,

FIG. 51.—Perspective view of Traction Engine.

and this, coupled with the large fire-tube and wide funnel, maintains a good draught when the engine is at work. The fire-box is open at the bottom, except the space occupied by the lamp. On the boiler near the funnel, upon the left side, there is a water-plug inserted, a lever safety-valve, $\frac{1}{8}''$ diameter, is fixed in front of the cylinder, with a spring-balance pivoted to a bracket soldered on the boiler; the valve can be tightened with a nut in the usual way. The cylinder, slide-valves, and link-motion (see working drawings) must be made according to the directions for making these articles under their respective headings.

The **cylinder** is $\frac{3}{8}''$ bore, $\frac{3}{4}''$ stroke; it is fixed by a coupling-screw to the steam-pipe, and rests on the top of the boiler near the funnel. There is a top and bottom circular guide-bar, both of which are soldered at their free ends into a ring made of sheet brass, and which rests on a fork when in position. The piston is $\frac{1}{4}''$ broad, having a $\frac{1}{8}''$ groove for packing, with a $\frac{1}{16}''$ collar on each side. The steam-pipe, about $\frac{1}{8}''$ diameter, comes out upon one side of the boiler, where, after having a cock screwed on, it joins, by means of a coupling-screw, with the steam-pipe leading to the cylinder; the other end of the steam-pipe is bent up inside the boiler to take steam as dry as possible, and prevent priming.

A piece of sheet brass or tin, of the same length as the cylinder, may be moulded round the latter on the side opposite to the valve-spindle, then curved downwards and be bent so as to rest by a small collar upon the boiler, to which, instead of by studs, it might be attached by a little solder. This makes the cylinder look as if it was encased in a jacket, at least upon one side, and gives it additional support to that derived from the steam-pipe, to which it is attached by means of a coupling-screw or union, so that by loosening the solder and the coupling-screw, after disconnecting the connecting-rod and eccentric-rods from the link, and the latter from the lifting-links, the cylinder can be raised up, the blast-pipe brought out of the funnel, and then easily removed for purposes of packing the glands, etc. A $\frac{1}{8}''$ cock screwed into the steam-pipe below the coupling-screw controls the steam to the cylinder; a wire running along the side of the boiler, terminating in a crank handle (the regulator), opens and shuts this cock.

The end of the valve-spindle rod is forked, and embraces the slot-link, which is cut out of a piece of sheet brass. The slot is struck to the correct curve, then drilled and filed to outline (see page 139). A snug is left at one end for pivoting to the lifting-links; these lifting-links, instead of being pivoted to one end of the slot-link, could be pivoted by a short arm to a central snug formed between the eccentric-rod snugs, and in this way the link would be raised and lowered much more easily than when pivoted by one end, and the engine would tend to work with less friction. The use of the short arms, one on each side, is to keep the lifting-links in a lateral line with the slot, and raise the link at the same time from the centre. Both arms must be bent outwards, so as to keep the lifting-links clear of the ends of the valve-spindle and of the eccentric-rods when raised and lowered. The lifting-links are made of sheet-brass strips, pivoted by a fork at the lower end to the weigh-bar, which revolves in bearings on the boiler; a small crank or slotted lever with a pivot is fixed at one end of the weigh-bar, this raises and lowers the link, and is worked by a sheet-brass rod, $\frac{3}{16}''$ broad, pivoted to the reversing-lever on the foot-plate, which works in a racket. The eccentric-pulleys are turned out of solid brass rod, bored and soldered on to the crank-shaft; they are $\frac{5}{8}''$ diameter over all, including collars; one is set for forward and the other for backward motion, but neither of them have scarcely any lead. The straps are of sheet brass, and the rods of brass wire (see page 116). The brass ends of the rods that embrace the links are forked, and are screwed to the ends of the (wire) rods. This forked end is made out of a piece of brass, with a saw-cut into which the slot of the narrow link passes. Both the fork and the link is drilled, and a small pin passed through to form a joint, and this is fixed to the outside of the fork by a bit of solder: plenty of free play must be allowed to the link on the pin. The other eccentric-rod is attached in the same way to the other snug on the link. In such a small engine, very great care must be taken, when adjusting the eccentric-rods, to make certain that they are both of the same length, as well as the *proper* length, for a very slight difference in their respective lengths will prevent the engine from running well, or even at all in one direction, while it will go whirring round in the other, and the cause of this seems rather puzzling to discover at first. In all cases, when it can be done, the slide-valve should be properly set, with the valve-cover removed, the cylinder being kept firmly fixed in position; press the slide-valve up against the steam-chest, following the rules given in Chap. IV. Do this with the forward eccentric-rod, setting it properly; then reverse the engine, turn the fly-wheel by hand in the opposite direction, and do the same with the backward eccentric. When it is impossible to remove the valve-cover after the cylinder is fixed in position, as in the case of locomotives, the valve-spindle must, before adjusting the eccentric-rods, have filed very lightly upon it marks showing its relation to the ends of the stuffing-box, or some fixed mark, when one port is full open, and when the other port is full open, and these file-marks are the *guides* when adjusting the lengths of the eccentric-rods. A small plug-hole leads into the valve-case, which is closed with a screw-plug; this serves for oiling purposes and for blowing out water from the cylinder before starting, or if the boiler should prime, as it may do if filled too full of water. The links are pivoted together by very fine rivets, made with a file out of thin copper wire; a head is left at one end, and the other end is put through the holes in the links, cut short, and riveted with a light hammer against the vice or an iron block; the superfluous part of the rivet-head can be filed off to look neat. If hammered tight, the rivet makes the parts quite a fixture (there is no joint); if hammered only so tight as to leave a little play, the parts joined form good workable joints for the link motion. If one lifting-link only is fixed on, this must pivot freely on the end of the short arm by which it is attached to the slot-link, while the other end of this arm is riveted immovably to the central snug of the slot-link. If there are two lifting-links, these pivot on a wire that passes through the upper end of the slot-link, and are retained on the wire by outside nuts.

The **fly-wheel** is 3″ diameter, of brass, screwed on to the built-up crank-shaft (see page 105). The axle revolves in solid bearings, fixed in holes slotted out in the side-plates. The axle is small, so instead of turning it down narrower at the journals, solder on two brass rings, in such a position that one lies inside the one bearing, and the other inside the other bearing; these prevent the axle from shifting. The rings are made out of sheet brass, and slipped over the ends of the axle. Each bearing is nearly $\frac{1}{16}''$ thick (the thinner they are the more easily does the engine work), and has a hole drilled through it for the axle to pass, with a small vertical hole leading to the journal for oil.

The **road-wheels.** The hind ones are $4\frac{1}{8}''$ diameter, and $1''$ broad; the rim of each wheel is made out of sheet brass, $13\frac{1}{8}'' \times 1''$, bent circular and soldered inside, at the overlap. Two collars cut out of tin, about $\frac{3}{16}''$ broad, are soldered all round the rim; one collar goes inside, near the outer edge of the rim, and the other collar is placed at an equal distance from the inner edge of the rim, and these are for attaching the spokes to. To make a collar, take a piece of sheet tin. From the centre with compasses describe a circle, say $4''$ diameter, then make a narrower circle inside, $3\frac{5}{8}''$ diameter; cut the tin to the outline of the outer circle, then cut to the outline of the inner circle; the narrow strip between the circles must be hammered and bent straight: it is best to halve it. Place this collar inside the rim (as mentioned above), and solder all round; if the ends overlap, they must be soldered together. Four collars are required, two for each wheel. A number of small narrow strips of tin can be soldered in a slanting direction at intervals around the rim of the wheels, to represent the cross steel plates in a real traction engine. The nave of the wheel is composed of a central boss, 1″ long, made of the finest brass tubing, and two central discs, one outside and the other inside, to which the spokes are attached; the wire axle passes through the brass tube in both wheels. Upon one of the wheels the brass tube will be about 1½″ long, and is left so as to project on one side of the wheel, and a tooth-wheel is firmly soldered to this projecting boss, so that it may gear with the other tooth-wheels; a pulley is soldered to the inner edge of the projecting boss of the opposite wheel for the brake arrangement, the wire axle passing through both pulley and tooth-wheel on opposite sides. Both wheels are loose on their axles, and are kept in position from coming off by means of small nuts screwed on the ends of the axles.

To make the central disc and the spokes, take a piece of tin, describe a circle 6″ diameter, then a smaller one from the same centre, $\frac{3}{4}$″ diameter, mark out between the circles eight radii, or spokes, each $\frac{1}{8}$″ wide, cut the tin circular to outline, then cut out the pieces of tin between each spoke up to the $\frac{3}{4}$″ circle. Now we have a central disc, and eight spokes radiating from it; four of these central discs and spokes are required, two for each wheel. Each disc is bored in the centre, and soldered at one edge on to the central boss; a space of about $\frac{7}{8}$″ is left between them, and they are fixed so that the spokes on the one side alternate with those on the other. It will be seen that in each wheel the spokes are sixteen in number, and that all the outside spokes are bent inwards, then cut short, and soldered to the collar upon the inside of the wheel; those again on the inside of the wheel are bent in the same way, cut short, and soldered to the collar on the outside of the wheel; thus the spokes or arms both alternate, and cross each other at the same time, as in a real traction-engine wheel, except that instead of rivets to fasten them we use solder. After fixing a tooth-wheel to the projecting boss on the inside of one wheel, and a small pulley to the projecting boss on the inside of the other wheel, both wheels are complete, and may be laid aside till the rest of the engine is completed. Washers, which can be made of pieces of brass tubing, may be slipped over the wire axles, between the wheels and sides of the fire-box, to keep the wheels at the proper distance from the fire-box on the inside, and nuts can be screwed on to the ends of the axles on the outside. The wheels must have plenty of play, and be quite loose upon their axles.

The axle is made of iron wire; this passes through the central boss of both wheels, and is firmly screwed into a brass block soldered on each side of the outer fire-box. The road-wheels require care in constructing, and are troublesome to make, to get them truly circular, and the crossed arms properly attached to the outside and inside collars, but when finished they resemble ordinary traction-engine wheels.

The **front wheels** are similar to the last, but much smaller, $1\frac{7}{8}''$ diameter, $\frac{3}{8}''$ broad, each with ten spokes. There is only one collar inside, soldered to the middle of the rim. The arms alternate and cross each other, as in the hind wheels. The wheels are fixed loosely on their axles, and are kept in position by the sides of the frame on the inside, and by nuts on the outside: they are provided with radial gear. The frame is made of No. 20 sheet brass, with an angle-plate projecting behind and at both sides, through which the axle passes. A raised disc is soldered on to the top of the frame, midway between the wheels, and this disc is pivoted to and rests upon another disc, riveted by means of two projecting flanges to the bottom of the smoke-box. The direction of the wheels is controlled by two small chains, attached to eye-bolts, which are fixed near each end of the back plate of the radial-gear frame. These chains pass one on each side over, and are fixed by small bolts to a brass roller—made of wire—revolving in two bearings, soldered one at each side of the front of the fire-box; one chain goes over, the other under the roller, and are wound once or twice round it, before being fixed by their ends to the bolts which pass diametrically through the roller. On one side the roller has keyed or soldered to it a bevelled tooth-wheel, which gears with another spur-wheel, fixed to a shaft supported in bearings above and below; this shaft passes along to the foot-plate, and ends in a hand-steering wheel placed at the driver's left hand. Turning this wheel with your hand, when the engine is running in either direction, winds one chain round the roller, tightening it, and unwinds the other chain at the same time, and so causes the radial frame to revolve on the central pivot—which consists of a bolt passing through the centre of each disc. This makes the engine run forward in a straight line when both chains are kept equally tight on the roller, or go to either side, according to whichever chain is the tightest. The tie-rod is a brass rod running in the median line, from the front of the fire-box to the centre of the back angle-plate of the radial frame. This is quite unnecessary, but renders the engine complete.

Tooth-wheel gearing. It is unnecessary to state the sizes of the wheels and pinions, as that can be ascertained by referring to the drawings, and also to mention the number of teeth in the various pinions and wheels, for so long as the wheels and pinions do not lock when revolving, the gearing will work perfectly: it only connects with one road-wheel. This wheel carries a tooth-wheel fixed on the inside projecting boss, and connects by means of the other tooth-wheel and pinions with the fly-wheel shaft. When it is desired to make the engine travel, this object is attained by the upper pinion: this pinion, by means of a lever pivoted to a bracket projecting out from the side frame, and working in a curved fork at its other end, can be thrown into gear with the road-wheel by slipping the pinion over the end of the crank-shaft, and letting the slot in the boss come in contact with a feather soldered on to the end of the crank-shaft (see Sheet No. 8). This lever carries a ring on its under surface, pivoted to it, and through this ring the outside projecting boss of the pinion passes, and revolves inside. This is retained in position by a collar soldered on to the end of the boss, and when fixed, a pin passes through the fork and the lever to steady it. When the pinion is on the crank-axle the feather enters the slot in the boss, and the pinion revolves with the axle, and this carries all the other wheels and pinions, as well as the road-wheel, round with it. The pinion is thrown out of gear by removing the pin out of the fork, and pushing the lever outwards, when the pinion will slip off the end of the crank-shaft. When *in situ*, the upper pinion gears into a tooth-wheel fixed on an axle, which carries a pinion placed outside of itself: this latter pinion gears direct into the tooth-wheel fixed on the axle of the road-wheel. It will be noticed that the intermediate tooth-wheel and pinion revolve together upon the same axle, supported in bearings contained in a sheet-brass frame of No. 20 metal, attached to the sides of the fire-box. This frame is open above and below, but for appearance' sake a tin cover is put over the tooth-wheels.

All the tooth-wheels and pinions required may be obtained from any watchmaker. They must be fixed up to revolve in the frame upon the engine, so that they do not lock when the road-wheels rest on the ground, and the engine is set to work. The above is not the arrangement of the gearing in a real engine, but it is easily fitted up, and works well.

The **tender** and **spirit-lamp.** The tender is small, of a width a little more than that of the fire-box. It had best be cut out of tin plate (tin from an old biscuit-box does well where a thin plate is required) to the correct size, then bent over a block by hammering to shape it out. The tender, or bunker, is composed of a floor, a back, and two sides. These last should have each in front a flange made to project inwards, so that when the tender is fitted to the engine, the flanges project inwards against the sides of the fire-box, and are fastened to it by solder. Along the upper border of the back, and each of the sides, solder on a coping. This is made of a single strip of tin, or separate pieces are joined together at the angles. After being bent to the proper shape, solder an iron wire around the tender to rest on the upper edges of the sides and back, as this, combined with the coping, gives a good appearance when finished. The foot-plate is cut out with the tender from the same sheet, or else is soldered all round to the lower edges of the back and sides. On one side two brackets are soldered for the brake-rod, and two steps are cut out of sheet brass and fixed on, as in a real engine. The draw-bar consists of two sheet-brass angles, soldered behind the tender close to each other, with a bolt passing through a hole drilled in the centres of both of them.

The **lamp** is made of tin, as described on page 51. The tank is rectangular, of the same length and width as the tender, and it can be made square or circular at the back: the latter is the best (see working drawings), as then, with the filler, it represents the water-tank in a real engine. The sides of the lamp, as well as the back, must be prolonged upwards to meet the tender under the foot-plate, and the top border of each side must be bent inwards for about $\frac{1}{8}''$: this is to enable the lamp to be drawn out and in, along a horizontal groove under the foot-plate (see working drawing sheets). The groove is made by soldering on both sides, under the full length of the foot-plate, a piece of cranked tin, so that the internal collar of the sides of the lamp passes in between the foot-plate and this cranked piece of tin. In order to be understood, we have exaggerated all the angles shown in the drawings of the lamp. The funnel for pouring in methylated spirits to the lamp is fixed at the side of the lamp and the tender. It is D-shaped in section, with the flat part close to the tender, and the convex part bulging outwards. It can be fitted on the top with a lid, hinged to it, and in this way it represents the inlet to the tank in a real engine. The steering-wheel is of brass, fixed on a wire shaft, that carries a bevel-pinion at its other end. This rod passes through two bearings, simply pieces of sheet brass soldered to the sides of the fire-box; a small brass ring is soldered above and below the bearings, to keep the pinion at the end in contact with the other one.

The **brake** (see working drawings) is entirely for appearance' sake, and to give a finish. It is placed on one side of the tender only. The brake-rod (of iron wire) passes through two stiff sheet-brass brackets, soldered on the side of the tender. The rod is kept from vertical movement by the brake handle, which is firmly screwed and soldered on the top of the rod, and a collar is soldered on underneath the upper bracket. A lever passes along from this rod, at right angles to it, and operates on the brake by tightening a brass strap up against a pulley which revolves with the road-wheel, and is fixed to its central boss. This lever is again acted upon by a pivoted nut, through which the brake-rod passes. In this nut, the vertical hole for the rod and the horizontal one for the pivots should be drilled in line with each other, but in a small size there is difficulty in getting a sufficient amount of metal on the sides of the central hole for the pivots to grip, so that it is better to drill the vertical hole either in front or behind the horizontal hole. In this way the nut will act just the same in operating on the brake, and the pivot can be formed by a copper rivet passing through the forks and the nut, and riveted on both sides: the vertical hole must be tapped by a suitable thread to fit the screw on the brake-rod. This screw, though capable of being freely turned by the handle, cannot lift, and will therefore raise or depress the nut and its lever. The strap is a strip of sheet brass pivoted at one end of a fulcrum, and passing about half-way round the pulley before being embraced by the end of the lever; this strap can be drawn tight on the pulley, or left slack, according to the direction in which the brake handle is turned. The brake might be connected with both road-wheels, but this would involve more labour, and its attachment to one wheel is quite sufficient.

Cotton wicks do for the lamp, but wicks of asbestos twine give better results, for if the spirits of wine falls short, this substance does not burn away like cotton, and on pouring in more spirits, the fire soon gets brisk again. The above is a description with drawings of an engine as made by the author, a short account of which he gave in the *English Mechanic* some years ago. The engine in working order weighs $3\frac{3}{4}$ lbs. fully. It will travel slowly backwards or forwards along a smooth floor, and haul a load; it will also mount a slight incline, consisting of a plank a little inclined at one end. When at work, the fire-door should be left open for admission of air to the lamp. We omitted to state that the side-plates which carry the crank-shaft bearings are made of sheet brass, one fixed on each side (by solder) to the fire-box. There is a tin cross plate between them in front, which fits the curvature of the boiler underneath, and which contains two slots, one for the connecting-rod to pass through, and the other for the eccentric-rods: this is called the "spectacle plate." It may be omitted, as it is troublesome to fit up.

Portable engine (see Fig. 52 and Sheet No. 9). This form of engine will not detain us long in describing, as it resembles the above, but the cylinder is set over the fire-box, and not at the smoke-box end of the boiler. There is no link motion and tooth-wheel gearing to fit up. As the engine is stationary when at work, the road-wheels can be turned and finished on the lathe from brass castings: their weight does not matter. The front wheels are attached to a pivoted frame below the smoke-box, as in the traction engine. The cylinder is set near the fire-box end of the boiler, and the fly-wheel at the smoke-box end. The brass bearings for the crank-shaft are bolted by their feet to stiff sheet-brass brackets, soldered to the sides of the boiler, as is done by some of the makers of portable engines. The safety-valve, $\frac{1}{8}''$ diameter, is fixed on the boiler at the side of the cylinder slide-valve case,

FIG. 52.—Perspective view of Portable Engine.

and connects with a small spring-balance pivoted to a bracket, which is soldered or screwed to the back of the boiler. A cock inserted between the steam-pipe and cylinder does duty as a regulator. The mouth of the steam-pipe inside the boiler should be turned up slightly, as in the traction engine. There is a screw-plug through which the boiler is filled with water; the exhaust-pipe comes out underneath the cylinder, and is led along the top of the boiler, and inserted into the base or lower part of the funnel: the mouth of the blast-pipe here, as well as in all the other engines, should be flattened a little with a hammer before fixing it in position, as this contracts the opening, and causes a stronger blast or rush of steam through the funnel, thereby intensifying the draught. If the pipe be too much contracted at the mouth, this tends to increase the back pressure in the cylinder, and prevents the engine from working so well. A hole should be drilled leading into the slide-valve case, which can be made steam-tight by a small cock or screw-plug, for the purpose of oiling the parts inside and blowing out the water at the start, or if the engine primes. This is used instead of putting on cylinder cocks, which are difficult to fit, being so small. There is only one eccentric, which is set to run forwards.

The lamp is made of tin, rectangular in shape, supporting eight or more wick-holders: there is a funnel through which to pour in spirits of wine. The lamp passes inside the fire-box, and rests on two long narrow sheet-brass strips, one near each end: these pass across from one side of the fire-box to the other, and are bent up outside at right angles, and riveted to the sides of the fire-box below the water-line.

The funnel can be made in one portion, or in two separate pieces, and hinged together. It may be made of brass tube, tin, or sheet brass. This last can be hammered over a circular piece of wood till it forms a tube; the longitudinal seam may then be soldered, and a rivet put through at the top and bottom: this will hold it firmly together. The lower part or base may be a brass casting, or be made of sheet brass: then it ought to have a collar turned down upon it, and be fixed by two rivets over a suitable aperture on the smoke-box.

For details of chimney hinge, refer to Sheet No. 9. Cut out two pieces of sheet brass to size and shape shown (see Fig. 1), slot out one of the pieces, so that A will fit inside space B quite easily, bend each piece to the same curvature as the funnel (see Fig. 2); the projecting or narrow portions A and B will now stand pointing upwards and downwards, when the free edges of the semi-circular portions are placed against each other. With pliers bend both A and sides of B at right angles to the semi-curved portions, fix one of the pieces to the back of the root of the long portion of the funnel, by slipping a small ring over it and the funnel (a picture ring will do), solder the parts together (the ring prevents the parts getting loose if the heat melts the solder), attach the other piece to the upper part of the lower portion of the funnel with a small ring in the same way, and solder them together; each portion of the funnel has half of the hinge attached to it. To fasten them together, with pliers bend the end of A into a circle, and do the same with the edges of B: see that they fall in line with each other, and A will lie inside B. Now push a small bolt through B and A (see Figs. 4 and 5), with a head on one side and a nut on the other, and the hinge is complete. If a small strip of sheet brass, about ¼″ deep, is bent circular and riveted together, so as just to pass inside and be a tight fit within the lower portion of the funnel, on raising the upper portion to the vertical position it will slip over the collar and be kept quite steady. The hinge should be so fixed that the funnel will hinge backwards and rest on a brass fork (a casting) screwed into the valve-chest. There is no cap on the funnel, which all portable engines have. This is omitted, as small boilers do not steam well with one, for it contracts the orifice of the funnel, and the fire burns badly. The outer fire-box and boiler barrel is made of No. 24 sheet copper, the internal fire-box, fire-tube, and tube-plate of No. 21 sheet copper.

Semi-portable or **underneath engine** (see Fig. 53 and Sheet No. 10). This engine is made by placing a locomotive boiler on the top of a flat-bed engine; a support or saddle-plate is fixed above the cylinder for the smoke-box to rest upon. The bed-plate is a brass casting, or is made out of No. 20 sheet brass, and riveted together. In real engines, the cylinder, steam-chest, stop-valve chamber, and saddle-plate are combined in one casting. As this would involve special patterns, and be troublesome in a model, we take the steam by means of an outside pipe leading to and screwed direct into the valve-casing, the exit from the boiler being controlled by a cock or a stop-valve. The steam-pipe

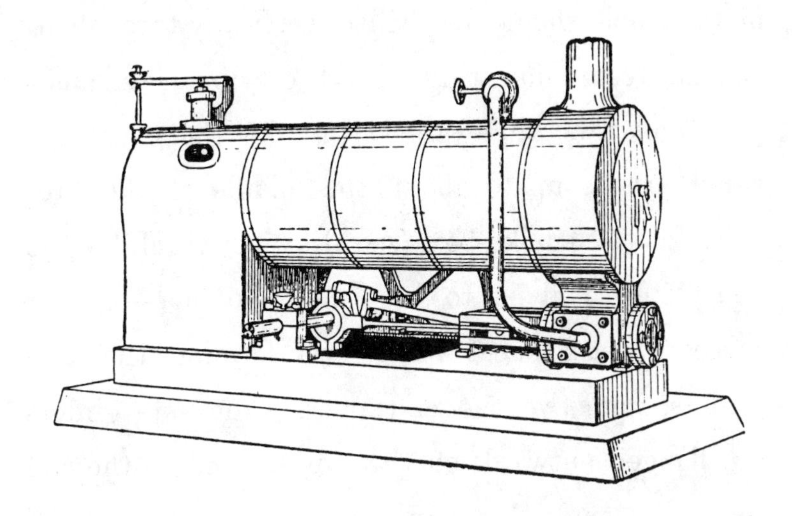

Perspective view of Semi portable Engine.

FIG. 53.

inside the boiler (see Sheet No. 10) is bent at right angles, its mouth being plugged, and a number of holes drilled in the upper surface for steam to enter, and in this way dry steam is got, instead of using a dome. The saddle-plate is a brass casting, either cast on the cylinder or, if separate, firmly soldered all round to it and to the valve-casing. The under surface is flat, the upper surface is cast with the same curvature as the bottom of the smoke-box, and supports it, this last being attached to it at the sides by small bolts and nuts. Before fixing the plate, a hole should be drilled vertically through it, one end communicating with the cylinder exhaust-port, and the other tapped to take the blast-pipe which passes up to about the middle of the fire-tube in the smoke-box (see drawings). Make the bed-plate—if it is to be a built-up one—of four pieces of stiff sheet brass, two forming the side girders, and two the end ones. The side girders are flanged outwards on the top, secured to the end pieces by angle brass inside, and riveted together perfectly square. At the foot, a piece of angle brass is soldered or riveted all round to the frame, except upon one side, for a short distance, to give room for the fly-wheel to pass through. The bed-plate is fixed by screw-nails, that pass through the flange to a wooden block or stand. Two sheet-brass cross-pieces, each of the same depth, are riveted inside between the frames (see drawings), to act as stays.

No longitudinal central girder is required, for the frame will be stiff enough without. A thin brass plate passes across the girders from the smoke-box end to the first cross-stay, and this is soldered or riveted to the side flanges: the cylinder and guide-bars are firmly bolted down to this plate. The guide-bars are four in number, fitted up in the same way as for the horizontal engine. The two bearings for the crank-shaft are placed near the fire-box end of the boiler (see drawings), and are bolted to the flanges of the side frames or girders. The fire-box is open underneath, rests upon and is fastened by means of angle brass and bolts to the flanges of the side frames and cross-piece at the back of the fire-box. The boiler must be raised so high that the crank and connecting-rod may keep clear of it when revolving. There is a spring-balance safety-valve on the boiler, and a plug for filling it with water. There is only one eccentric, set for forward motion. The wooden frame must be slotted out for the fly-wheel to pass through on one side. The outer fire-box and barrel is made of No. 24 sheet copper, the internal fire-box, fire-tube, and tube-plate of No. 21 sheet copper, and frames of No. 20 sheet brass.

A more powerful working engine, and one more easily fitted up, can be made by doubling the dimensions over all, or nearly so, of that on Sheet No. 10. Make the boiler, the same as described on page 46, of No. 16 sheet brass and No. 18 sheet copper: all the joints must be brazed. The cylinder is 1″ bore, 2″ stroke, which along with the saddle-plate, fly-wheel, and bed-plate may be made from brass or iron castings; but brass, being softer than iron, is the easiest of the two metals to work, and is not liable to crack when being drilled, like iron. The bed-plate may be a casting, fully $\frac{3}{16}$″ in thickness, or built up of longitudinal girders composed of No. 16 sheet brass: a slot must be cut out for the connecting rod "big end" and crank to pass when revolving. The boiler rests at the fire-box end, and is attached by angle brasses; screws fasten the bed-plate to a wooden sole-plate. The force-pump described on page 71 will be suitable for this engine. And if a small gas-jet is burned inside the fire-box, the engine will work away for an hour or two at a time without much attention, if the pump is regulated so that a constant water-level is maintained in the boiler. The bearings are entirely of brass. The crank-shaft must be a forging. A $\frac{3}{16}''$ cock will be large enough to supply the cylinder with steam, through a pipe of the same bore.

CHAPTER VIII.

LOCOMOTIVE ENGINE AND TENDER.

WE will describe in detail the making of a locomotive engine, in such a manner that little difficulty need be experienced in the construction of any of the several forms illustrated in this work. There is considerably more labour involved in a locomotive than in any of the engines hitherto mentioned. It includes a boiler, with a pair of cylinders, fitted with reversing-gear, and all the motion work connected with the engines must be carefully fitted and adjusted, so as to work satisfactorily under steam when made. The locomotives we propose taking up, though perhaps not just "correct models" in every minute detail, will be found to be exact semblances to the real thing, work well under steam, and be superior to any of the cheaper slide-valve locomotives sold in the shops, approaching in their construction the "powerful models of modern locomotives" as sold by Mr. Lee and others, and which from their high prices are beyond the reach of the ordinary amateur. The construction of any of the locomotives may be undertaken by an amateur of ordinary ability, without heavy or costly tools, and with every reasonable prospect of accomplishing it in a satisfactory way, provided he has patience to carry out the task to the end. These small locomotives will require accurate workmanship to ensure success, but the work is very light, and when finished, if carefully made, they will run well upon rails. If any one would like to construct a more powerful locomotive, let him refer to page 239, and to Sheet No. 13; these working drawings are so proportioned, that by doubling the dimensions over all, or nearly so, a large and powerful model can be made.

All the locomotives have coupled driving-wheels, except in one instance: they have single frames (with or without bogie), and outside cylinders, set parallel with the axis of the boiler. We have adopted outside cylinders, as inside cylinders and a crank-axle are difficult to fit up, for want of room, between the frames; and the axle, being small, must be a "built-up" one, and probably would not last very long if run often.

This locomotive (see Sheet No. 11, A, B, C) is called a bogie locomotive, as the front part of the frame and smoke-box rests upon a four-wheel truck. All the dimensions can be got by referring to the working drawings.

The **frames** are made from No. 20 sheet brass, and form the attachment for the fittings. They receive and sustain all the stresses due to the moving parts; they are supported on the journals of the axles, and carry the boiler, with the cylinders and valve-gear. The two side frames, or frame-plates, are made of stiff sheet brass rather than a casting, to save weight, mark and cut to outline (see drawings). Both frames are attached in front to a wooden buffer-beam A, by means of sheet-brass angles, on one side only, riveted or soldered to the frames by copper wire rivets, and attached to the beam by very small bolts and nuts. At the back they are held together by a transverse sheet-brass plate B, in length equal to the buffer-beam. Care must be taken that the side frames and the end ones are at right angles to each other. To stiffen the frame, two cross-stretchers, C, D, are riveted inside by angles, one passing in front of the fire-box, and the other behind the cylinders. This last is called the "motion-plate," and to it the slide-valve spindle-guides are fixed in a real engine; but here this plate is simply slotted out near both ends, to allow the slide-valve spindles to pass through. A drag-plate casting at the hinder end is omitted to save weight, as well as axle-boxes, horn-plates, and springs. The straight axles simply pass through holes drilled in the side-plates to receive them; the rectangular holes slotted out in the front part of the frames on each side, over the bogie-truck, are to take the steam-chests of the cylinders, which are thrust through them, the cylinders being outside the frames, and the valve-chests inside. Underneath the cylinders a transverse stay passes from one frame to the other, and is attached to each of them by rivets; the under surface of this stay forms the support for the bogie-truck, which rests against it; a pin passes through the centre of the stay and the bogie-truck, on this the latter pivots; a nut on the under surface of the bogie-pin completes the attachment. The buffers are all four of brass, turned on the lathe; the front ones, corresponding with those on the back of the tender, are larger than those on the hind frame of the engine. Each of them is centred, tapped with a thread, and a screw put through, by which they are attached to the frames. The screws pass through the buffer-beam and hind frame, and tighten up the buffers with nuts. It will thus be seen that the buffers are solid, and have no springs.

The **bogie** consists of a rectangular frame made of No. 20 sheet brass, flattened by hammering, and riveted together. There is a central pivot-hole, drilled in a transverse stay, riveted on to the top of the frame, through which the central bolt passes. The holes for the axles are simply drilled through the sides of the frames, at a distance of $2\frac{3}{4}''$ from each other; there are no springs. The cow-catchers are made of sheet brass, and riveted on to the sides in front, so as to project before the wheels, and keep clear of the rails.

The **wheels.** Those of the bogie are four in number, of cast brass, turned and faced on the lathe. They are $1\frac{3}{16}''$ diameter on the tread, and are screwed firmly upon straight axles, which have been previously turned in the lathe. They, and all the other wheels, must be bored straight through the centres, and fixed so as not to wobble, but revolve truly on their axles. The driving-wheels, four in number, are made from castings. It is necessary to turn the wheels on the lathe, and this can be done by drilling them, after correctly centering, and mounting them on a temporary shaft between the lathe-centres: turn them on the treads, flanges, rim, boss faces, and edges. All the wheels should be slightly coned on the tread, and the flanges of them all thinned down on the lathe to just over $\frac{1}{32}''$ in thickness, as this diminishes friction. The spokes can be filed smoothly over, and painted afterwards. The axles for these wheels are straight ones, made of iron wire fully $\frac{1}{8}''$ thick, cut to the correct length, tapped with a screw, and the wheels firmly screwed on to them. It is a good plan to turn the ends of the axles where they enter the boss of the wheels to a smaller diameter, as in this way a collar is left, upon which the wheels can be screwed up tight, and are likely to remain so. Without this arrangement, especially where reversing-gear is put on and the wheels run in both directions, they are apt to shift on their axles, from the thread of the screw getting slack, which must be prevented from occurring. The bogie-wheels can be fixed in any position on their axles, but the driving-wheels must be firmly fastened, so that the crank of the one wheel is set at right angles to the other crank upon the same axle. The driving-wheels could be fixed firmly on their axles by means of a small set-screw, a hole for which has been drilled upon the inside projecting boss of each wheel, running at right angles to the central hole for the axle. This plan does well if the screw is made fairly strong, and it can be tightened up with pliers after the wheel is on the axle.

The **coupling-rods** pass between the driving-wheels. They are made of soft iron, and filed to outline $\frac{1}{16}''$ thick and $4\frac{1}{16}''$ between their centres; they are not bushed with brass, the holes for the cranks being simply drilled through them. These rods rest upon plain crank-pins screwed into the boss of the wheels, and are kept on by nuts which screw up against a collar on each pin. This collar is made by making the end of the pin which enters the nut a size smaller than the body of the pin, and tapping it with a thread. The nut could be soldered on, but this is not handy, if it be required to take the rods off at any time. The crank-pin of the driving-wheel is the same, but longer, than that on the trailing-wheel, and passes through both the connecting-rod and the coupling-rod at the same time.

As some amateurs may prefer to have **axle-boxes** fitted on the frames, which are wanting in the above locomotive, we will mention how these can be applied, at least for the driving-wheels. Make the axle-boxes of undivided rectangular blocks of brass (see Sheet No. 11, C), to suit straight axles; these fit and slide between the horn-blocks B, and being flanged, slide upon the faces of the horn-blocks on one side, and of the frame-plates A on the other; the holes for the axles are drilled through the solid boxes. The horn-blocks are cut out of No. 20 sheet brass, of arched form, and are fastened to the frame-plates by small rivets or solder. The axle-boxes C must be fitted by filing to be an easy sliding fit between the faces of the horn-blocks and the frame-plates; file the outer faces of the horn-blocks smooth, to be an easy fit with the guide-strips of the axle-boxes. It is best, for small engines, to adopt spiral wire springs. The spring enters a shallow hole drilled in the top of each axle-box; it bears between the bottom of this hole and the lower face of the arched horn-block: small horn-stays pass across the foot of the horn-blocks, and are kept in place by screws or solder. To diminish friction, the axle-boxes must not be more than $\frac{1}{16}''$ in thickness, and if adopted, a wider gauge of line will be required than $2\frac{13}{16}''$.

The **cylinders** are fitted up as described on page 79. Each cylinder has only one circular guide-bar (of iron wire), screwed into the back cover at the top part, to save friction. The free end of the guide-bar is left loose, and is not attached to the frames, but it might be so fastened. The cross-head is forked, and has a long projecting guide, which embraces the guide-bar (see dotted lines in Sheet No. 11, A). It is a good plan to screw into each valve-chest a small cock through which oil can be poured, or even injected, to oil the slide-valve and piston each time before running the engine, as this tends to diminish friction, and the engine runs more easily. When in position, the steam-chests of each cylinder are connected together by means of a horizontal pipe passing between them, and they are so placed that the chests lie in the rectangular holes in the side frames, while the cylinders are outside; and if flanges are not cast on the cylinders for attaching them to the frames, these must be made of sheet brass, and soldered round the steam-chests outside the frames, and attached to the latter by bolts in order to give rigidity to the cylinders, and prevent them from rocking. Be careful, when fitting, that the centre of each cylinder is at the same distance from the centre of the driving-axle, and that the central line of the axis of the cylinders, if prolonged, would pass through the centres of the driving- and trailing-wheels, and also be at right angles to the axle, and quite parallel with the frames.

The **pistons** are of brass, $\frac{1}{4}''$ broad, having a $\frac{1}{8}''$ groove for packing, with a $\frac{1}{16}''$ collar on each side of the groove.

The **connecting-rods** are made of soft iron, and have solid ends; but if preferred, the "big ends" can be made as described on page 97.

Valve-gear. The slot-links are made of sheet brass, similar to that on the traction engine (see page 172). The lifting-links (of sheet brass) are two in number, one for each slot-link, placed upon the outside only. Each pivots upon a short arm, which is riveted immovably upon a central snug on the slot-link (see page 172), or these could be attached to a snug upon the lower end of the link in the line of curvature, and be fixed to it by a small bolt passing through both, and fastened with a nut. There are four eccentric-pulleys on the driving-axle, two set for forward and two for backward motion. The eccentric-straps can be made of sheet brass, the rods of brass wire, and the forked ends cut out of solid brass and screwed to the rods; the bolt on which the link pivots is fastened to them with solder. For their construction see pages 116, 137, and 139, also Sheets No. 11, C, and No. 8, C.

The **weigh-bar** passes across under the boiler from one frame to the other, being supported in sheet-brass bearings soldered to the frames. There is a small lever on each side to pivot to the lifting-links, and a rod passes along at the side of the boiler, inside the splashers, to a reversing-lever moving in a notched quadrant placed on the foot-plate (see longitudinal section of locomotive, Sheet No. 11, A). The reversing-lever and quadrant are placed on the left-hand side of the foot-plate. The travel of the slide-valve is $\frac{3}{16}''$, the steam-ports are $\frac{1}{16}'' + \frac{1}{4}''$, and the exhaust-port $\frac{3}{32}'' + \frac{1}{4}''$.

The **boiler** is made of No. 24 sheet copper, and all the dimensions will be found on Sheet No. 11, A.

The **internal fire-box** is made of No. 21 sheet copper, and must be brazed. For the method of making it, see page 37. We employ thinner copper for the outside of the boilers, in order to render them lighter than they would be if the whole was made of No. 21 sheet copper.

The **regulator** is a cock attached to the steam-pipe in the smoke-box after it emerges from the boiler. This cock is opened and shut by a small wire which passes through a $\frac{1}{16}''$ bore brass tube, soldered into each end of the boiler near the top. One end of this wire is pivoted to the regulator handle on the back of the boiler, which moves on a pivot in the ordinary way. When this handle is pulled out by one end, it opens the cock, admitting steam to the cylinders; and when pushed in close to the boiler, shuts the cock. The smoke-box end of the wire is bent downwards so as to form a kind of loop, which embraces the crank-shaped handle of the cock (see Sheet No. 11, B). The wire is of such a length that when the regulator is (shut) close to the boiler, the cock handle is pushed out near the smoke-box end and closed; and when the regulator is pulled out, the cock handle is drawn towards the end of the boiler and is then open. Putting this wire through a $\frac{1}{16}''$ bore tube saves making two stuffing-boxes at the ends of the boiler for it to pass through, as these, being small, are difficult to construct and keep steam-tight. A coupling-screw connects this cock to the vertical pipe underneath, that joins the horizontal pipe passing between the cylinders; the cock at its upper orifice is screwed to the steam-pipe which passes out through the tube-plate (the steam-pipe is soldered to the tube-plate). Of course a vertical opening for steam to pass through must be made in the horizontal tube, before the vertical one is soldered to it. Inside the boiler, the steam-pipe is carried along and passes up inside the dome; it is simply pushed through the tube-plate and soldered: there is no support required for the pipe inside the dome. There is only one steam-pipe, but two separate blast-pipes.

The **safety-valve** is made of brass, and instead of being fixed on a separate casing it is mounted upon the dome. This is a brass casting which must be turned on the lathe, and in order to chuck it the dome should be centre-punched at the top, driven tightly upon a turned piece of hard wood, then fixed between the lathe-centres, and turned down to the correct size. After this, drill two holes through the top, into which two brass pillars are fixed. These are made of pieces of brass tube screwed into the dome: one of them forms the fulcrum, and is slotted out for the lever to pass through—its orifice is plugged underneath—the other pillar forms the valve proper, and communicates with the interior of the dome. Upon it a valve-seat is formed, about $\frac{1}{8}''$ diameter, and a valve is turned to fit the seat. The lever, cut out of sheet brass, is pivoted to one pillar, and passes through a longitudinal groove in the top of the valve as it rests in the other pillar. A small spiral brass spring, which can be procured from any watchmaker, holds the lever down upon the valve, being so placed that its coils fill the space between an eye-bolt and the lever (the stem of which passes through the dome, enters a nut, and is soldered inside). One end of this spring is bent round and inserted into the small ring or eye-bolt, and the other into the eye of the lever; the fitting of this spring is rather difficult, but can be done with a little trouble. The valve has the appearance of the "Ramsbottom type," and if properly made it will act; however, we advise a direct-acting spring safety-valve to be attached to the boiler over and above, which can be relied upon to blow off excessive pressure of steam.

See Sheet No. 11, A, for back view of the locomotive with fittings. There is a fire-door, made of sheet brass, with hinges and a latch. These are riveted to the door, but soldered to the boiler. The regulator handle is pivoted horizontally to a piece of angle brass, soldered to the boiler on one side of the middle line, and on the other embraces the rod coming from the cock in the smoke-box, and is made of iron wire. There is no water-gauge, force-pump, or pressure-gauge attached. Two small cocks are soldered into the boiler to ascertain the height of the water; these should be purchased ready made. For description of reversing-gear see page 137, and Sheet No. 11, C.

The **cab** is made entirely of thin sheet tin. The top and the sides are in one piece; the saddle-plate arches over the back of the fire-box, and fits closely around it. The sides of the cab are bent circular with the top arching over the foot-plate, and are cut to a curved shape, so as to rest upon the top sheet of the hind splashers, and be soldered to them as well as to the saddle-plate. To strengthen the joints, three brass angles can be soldered on inside, between the cab sides and top, to the saddle-plate. A hand-rail pillar, made of wire, is soldered to the splashers and to the sides of the cab, and joins them by a strip of sheet brass which embraces the hand pillars. The windows are made by drilling two $\frac{9}{16}''$ holes in the saddle-plate, inserting glass discs—microscopic cover-glasses, $\frac{3}{4}''$ diameter, do well, and can be had from any optician. These discs rest against the edges of the window apertures in the saddle-plate, and around their edges a thin ring of tin plate, slightly overlapping, is soldered all round to keep them fixed. The foot-plate is the floor-space from the back of the framing up to the fire-box, and is fitted between the inside back splasher-sheets, to which it can be soldered. The splasher-sheets are made of tin, consisting of outside pieces and a top; there are no inside splashers, except over the back or inside half of the trailing-wheels, to conceal them from the foot-plate. Upon each side there is only a single splasher for the two wheels (see Sheet No. 11, B). This saves much work in fitting, and is adopted on the North-Eastern Railway. The splashers rest upon the side-plates of the engine: a piece is cut out on the under surface to let the coupling-rods pass without striking the sides; an angle piece, in section like **Z**, passes along the entire length at the foot, and by it the splashers are fastened with bolts and nuts or solder to the side-plates. The top sheet is of tin, fastened to the splasher sides with solder by one edge, and by its other rests against the boiler.

Front view of locomotive (see Sheet No. 11, A). The smoke-box door is circular in shape, made of tin plate, but not dished; small hinges are riveted to it and to the smoke-box. The door is secured with a bar and a T-headed bolt moved by the handle. The buffers are screwed into the buffer-beam, and secured by a nut.

The drawings (Sheet No. 11, A) show the smoke-box in section from the front. On the left hand the section passes through the steam-pipe, on the other through the exhaust-pipe. The steam-pipe is common to both cylinders. The single fire-tube is seen opening into the smoke-box, in front of which the steam-pipe, $\frac{1}{8}''$ bore, and the cock pass, being attached by means of a coupling-screw to the vertical pipe, which is soldered to the horizontal pipe passing between the cylinders. The object of having a coupling-screw in this pipe is for the purpose of being able to remove the boiler from the framing, if it is required to do so. The horizontal pipe is screwed or soldered into the steam-chests. The blast-pipes are screwed each into the openings of the exhaust-ports; they are bent inwards, and cut short so as to stand no higher than the middle of the fire-tube, as this height for their nozzles gives the best draught. Their mouths are flattened to contract their orifices, and so intensify the draught.

The **boiler** is attached to the frames by the coupling-screw at the smoke-box end, as mentioned above; and in front, on its under surface, the barrel rests on the concavity of a vertical piece of tin passing between the frames, and this forms the under surface of the back of the smoke-box (underneath the barrel). At the fire-box end the boiler rests on the side frames by brackets, which are not fastened to the frame-plates, and so allow of expansion: the brackets are soldered to the fire-box sides at a height to keep the boiler level. When the cab is firmly fixed, the boiler cannot shift.

The **smoke-box** should be made so that it will push along the front part of the frames over the bogie on to the boiler, and pass as far back as to rest against the plate underneath the barrel, and a wire can be soldered all round the barrel as a finish, to fill up the space between it and the boiler. The smoke-box must be cut away for about three-quarters of its length on the under surface, to allow of being slid past the steam- and exhaust-pipes, and to close this opening a small tin plate may be put across under the smoke-box, between the frames, resting on the steam-chests, as the cylinders are not contained in the smoke-box. The smoke-box is fastened to the frames by a small bolt and nut on each side. The funnel is made of tin riveted to the smoke-box, set over an aperture made for it, and a wire is soldered round the top. Sheet No. 11, A, gives the half-sections of the boiler as seen from the foot-plate and front buffer-beam ends, along with an elevation, longitudinal section, and plan. Sheet No. 11, B, gives plan of the engines, the frame-plates, and numerous detail drawings. Sheet No. 11, C, is devoted to the tender, lamp, brake, and link motion.

The **foot-plate,** with **side-plates,** had best be cut out of a single sheet of tin, and made so as to clear the boiler and driving-wheels when in position. It is attached by bolts to the angles on the top of the inside frames: it could be made in strips, cut out, and fitted all round outside. Angle-pieces had best be soldered on to each side of the projecting plates along their whole length, to give a better appearance and finish to the engine. The steps for the driver to get on to the foot-plate are cut out of sheet brass: they hang downwards, being soldered to the insides of the side frames.

The **spirit-lamp** is made of tin: the rectangular tank fits in under the foot-plate between the frames. It passes inside the fire-box, and rests upon two strips of brass riveted across near the front and back end: it carries twelve $\frac{3}{16}''$ wick-holders. A funnel is soldered into the end of the tank: this passes up inside the buffers (at the end of the locomotive) to about the height of the foot-plate, and spirits of wine is poured into this to feed the flames, through a corresponding hole cut in the flap cover-plate. When the tender is uncoupled from the engine, the lamp can be taken out as required.

Having described the locomotive in detail, we now take up the **tender**, and refer the reader to the drawings on Sheet No. 11, C. There is a longitudinal view, with a longitudinal section down the centre of the tender. The lower half of the plan drawing is a half view of tender as seen from above. The upper half contains three plan sections: at A, the top and sides of the tank are removed, and we see the floor-plate B; at C, the plate B is removed, and the view is taken on the angle D, which unites the framing and the floor-plate; at E, the section is below this, through the centre of the front wheel and axle-bearings. There is also a view of the front of the tender. The tender (to save weight) is made entirely of thin tin plate, and soldered together. The framing consists of two longitudinal frames cut to shape as shown; the wheels are inside, and the axles simply pass through holes drilled in the frames. The distance from the centre of the back wheel to the front one is 5″. The side frames are united at the ends, by means of angles, to two transverse stretchers: no intermediate bars are required, as the tender frame is stiff enough without them. On the upper edges of the side frames tin sheet angles are soldered, to carry the floor of the tank and coal-bunker, the wooden buffer-beam is attached to the hinder end (as in the locomotive) by the buffers, which are bolted to it and pass through the hind cross-stretcher as well, and are held in position by nuts. The tank is saddle-shaped, the sides are cut out of tin and bent to shape; they are soldered all round to the floor, and to the sides of the tender in front. The cover-plate of the tank is soldered to the sides of the tank and tender. The filler is a screw water-plug, the seating of which is soldered into the top of the tank, over a suitable hole at the back part. At the front end a tin plate crosses from one tank side to the other, its upper edge, as well as the hole through which the coal is drawn, can have a wire soldered all round, to give a good finish. Two steps of sheet brass can be soldered to the frame-plates on each side. The coping is made of tin, bent and mitred at the hind corners, curved in front, and soldered all round to the sides; a wire may also be soldered round the top to give finish, the hand pillars (of wire) are soldered into the plate at the bottom end, and to the curved support at the coping.

The wheels are $1\frac{1}{4}''$ diameter, placed inside the frames. The axles are prolonged through them, and when in position rest in holes drilled in the side frames, and box-shaped covers of tin are soldered on (over the ends of the axles) to the outside frames, in order to represent real axle-boxes. The springs are pivoted at the ends, and soldered to the side-plates. These are only imitation springs, and put on for appearance' sake. They may be made from brass castings, or separate narrow strips of sheet brass can be bound together by a brass strap and attached with solder, as is done in this case; a wire hanger passes down from the centre of each spring, and rests upon the corresponding axle-box. Two buffers, similar to those on the back of the locomotive, are screwed into the front plate of the tender. A hinged sheet-brass eye-bolt is fixed between them, through which a hook attached to the engine passes, when the locomotive and tender are coupled together.

The **flap cover-plate** F (see longitudinal section) covers over the open space between the tender and locomotive, to prevent the foot slipping between them. It is a strip of tin, flat or slightly curved, attached by a strap hinge to the tender foot-plate, and rests on the locomotive foot-plate when turned down. On one side there is a small hole drilled, corresponding with the position of the lamp-filler, and through this, when the locomotive is running, spirits of wine can be poured in to feed the wicks.

The **tool-boxes** are made of tin, each having a hinged lid, and are soldered on the top of the tank.

The **hook**, or draw-bar, is made out of a bit of wire, having a screwed shank which passes through the buffer-beam or hind frame, and is kept in position by a nut, a small brass collar having previously been soldered on it near the curve of the hook, to rest against the buffer-beam on the outside, when the nut is tightened up on the inside.

There is no connection for water between the tank and the engine, but for appearance' sake a small brass tube, $\frac{1}{8}''$ bore, may be inserted at one side into the tank underneath the tender and soldered, being previously bent so as to project forwards. A cock can be inserted in the forward part of this pipe, with a wire stem passing up through the foot-plate, and turning in a little bearing soldered to the front of the tender, which is opened and shut by a small handle. A corresponding pipe can be attached to the engine and appear to communicate with the boiler (see Sheet No. 11, A), and rest against the tender feed-pipe, being in line with it when the latter is coupled up; and a piece of thin india-rubber tubing can pass over the junction of the two pipes. One would suppose a coupling-screw might do, but this attachment is rather too rigid, and does not allow of sufficient play between the engine and tender. The pipe and cock may be entirely omitted.

The **brake** (see Sheet No. 11, C). The pillar is a casting, or is made of stiff sheet brass, soldered to the foot-plate at one side, having a cross-bar on the top, and through the central portion of the cross-bar a hole is drilled for the brake-rod A. This rod is kept from vertical movement by a collar B, soldered on it below the cross-bar, and by the boss of the hand-wheel firmly fixed to it on the upper surface. The rod carries a screw near its lower end, and by means of the hand-lever raises and lowers a pivoted nut C, attached to the end of a bell-crank lever D, similar to that described for the traction engine. The bell-crank is made of stiff sheet brass, and is fixed on a brake-shaft, the bearings for which are in two brackets soldered to the under surface of the frame of the tender; the lower end of the bell-crank is pivoted between the forks of the brake-rod, passing from the cross-shafts behind the front wheels. The remaining brake-rods correspond on the two sides, and the description of one serves for all. The hangers E are made of sheet brass, and are suspended inside the frames, and each has a horizontal brake-rod F, passing through to that on the opposite side. Each hanger is made of a strip of stiff sheet brass, having an eye at the top, which pivots in bearings soldered to the bottom of the tank floor. The brake-blocks G are made of hard wood, cut to suit the curvature of the wheel, and are fastened to the hangers by two pins of brass or iron wire, simply driven into the blocks through the hangers, which are amply sufficient to keep them tight. A horizontal rod H, pivoted to the lower ends of the hangers, connects all the brake-blocks together on one side. If this brake is well fitted, all the parts will move in unison on turning the brake handle, and it will act well.

A **screw-plug** (to pour water in) is screwed into the top of the boiler, and there is a whistle; the stem of the cock passes through the cab saddle-plate, and joins a handle above the boiler. On the left side of the boiler (see elevation), a small block of brass with a pipe leading down to the side frames is attached: this represents the feed-pipe and clack-valve box in a large locomotive.

There is no cleading around the boiler, *i. e.* strips of wood neatly fastened all round the boiler with brass hoops, to retain the heat, as this locomotive is rather small to have it. One or two sheet-brass rings may be soldered round the boiler to give a good finish.

Spring buffers can be made as follows. The buffer itself is turned circular out of a piece of brass rod, and consists of a head and a stem; the first is $\frac{1}{2}''$ diameter, the second is $\frac{1}{4}''$ diameter, and the whole is $\frac{3}{8}''$ long. In order to make plenty of room to get in a spring, drill a central hole fully $\frac{1}{16}''$ diameter, from the narrow end up to near the head, screw into the far end of this hole a wire spindle which has been filed to a narrower diameter all along the stem, slip a brass spring over this stem, of a size that will move freely in the central hole. Spring and wire must be then cut to the proper length. The buffer stem fits inside (with a telescopic motion) a turned brass casing, $\frac{7}{16}''$ diameter and $\frac{3}{8}''$ long, which is bored out to receive it, and is provided with a flange at one end, for attaching by two bolts and nuts to the buffer-beam. A central hole, scarcely $\frac{1}{16}''$ diameter, must pass through the flanged end of the casing, and communicate with the central aperture in the buffer. The pin that passes from the buffer goes through this hole when the buffer is placed inside the casing. The head of the spindle, being outside, moves backwards and forwards (in a wide, horizontal hole drilled to receive it through the buffer-beam) in response to the movements of the buffer against the spring in the casing. The head of the spindle coming in contact with the flanged end prevents the buffer from projecting too far out, or even being pulled out. The stem can be easily removed, when the buffer is unfastened from the buffer-beam, by simply unscrewing the spindle. The above plan is not the way the buffers are made in actual practice, but does very well, and is adopted as the easiest way of making spring buffers when they are so small.

The above is a description of a good working locomotive and tender, which was designed and made by the author, and from this engine all the working drawings were taken.

CHAPTER IX.

LOCOMOTIVE ENGINES.

Bogie-tank locomotive (see Fig. 54 and Sheet No. 12). A locomotive of this kind, having no tender to pull, will run along very easily. The method of its construction is similar to that of the last one, and being so we will not enter much into detail, but take up points on which it differs from that engine.

The frames are made in the same way, but are longer, and are prolonged behind the foot-plate in order to carry the coal-bunker. This is made similar to the tender, with a coal space, separated by a cross-plate from the foot-plate, with an opening for shovelling out the coals in the centre. The driving-wheels are concealed from view above the side frames by the side tanks, which are only imitations and not real tanks for holding water: both are made of tin. Each consists of an outer plate, which by means of angles is bolted to the side frames; the ends (the front one fits the convexity of the boiler) and the top are soldered to the outer plate, the inner edges rest against the boiler barrel. An inside plate only extends from

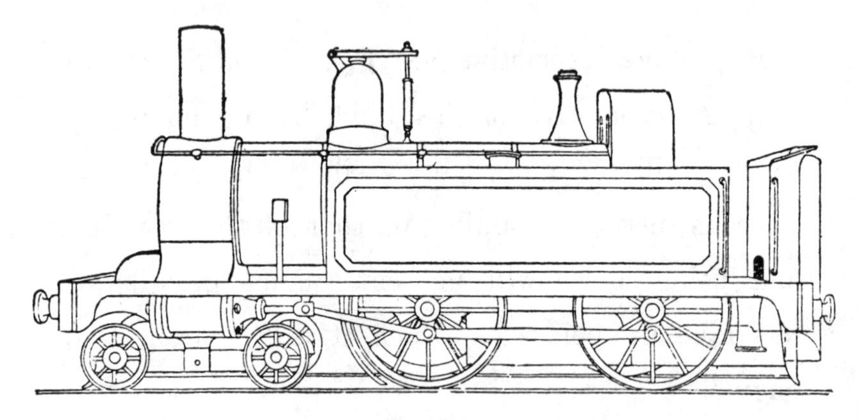

FIG. 54.

the hind end up to the fire-box on the foot-plate at both sides. The sides of the cab are soldered to the top of the tanks, fitting over the saddle-plate in front.

The lamp is made of tin, and this slides along a groove underneath the coal-bunker and foot-plate, between the wheel frames; the neck, which passes into the fire-box, contains fourteen $\frac{3}{16}''$ wick-holders. This lamp is similar to the last, but the funnel—for pouring in spirits of wine—comes up at the back behind the buffer-beam, and is placed inside the right-hand buffer, passing as far up as the coping of the coal-bunker.

The bogie-truck is similar to the last, except that instead of sheet-brass cross-pieces, the front, and back stays consist of copper wires riveted to, and passing between the side frames. The bogie-wheels are not solid, but have spokes. There is no link reversing-gear, but reversing is effected by loose eccentrics, one for each cylinder, similar to that described on page 134.

The safety-valve is placed on the top of the dome, the crown of the dome being its seating. A boss must be cast on the dome, and drilled to receive the valve. A lug or fulcrum is screwed into the crown to take one end of the lever (pivoted to it). A spring-balance, attached by a pivoted joint to the top of the boiler, has its rod passing through the other end of the lever, and the lever is tightened on the valve by a nut. This valve does act, but is more for appearance' sake. The direct-acting spring-valve in front of the cab is the one that relieves the boiler of excessive pressure, and must be so arranged by tightening the spring that it will blow off when the pressure rises above 30 lbs. per square inch, but not under that. The valve screws into a casing, soldered on the boiler, and through this we pour water into the boiler, after removing the valve for this purpose. There is no whistle, or brake applied, but we will describe one suitable for this locomotive.

The boiler is made of copper, the dimensions being the same as that described on page 37, to which reference can be made. Cylinders are $\frac{7}{16}'' \times 1''$, steam-ports $\frac{1}{16}'' \times \frac{3}{8}''$, exhaust-port $\frac{3}{32}'' \times \frac{3}{8}''$. The barrel and outer fire-box is made from No. 24 sheet copper, the internal fire-box, fire-tube, and tube-plate of No. 21 sheet copper; the frames are made of No. 20 sheet brass.

Details for the construction of a link motion, if it be desired to fit one on, can be got by referring to pages 136, 139, and 171, but if the locomotive is fitted with loose eccentric reversing-gear, it will run more easily under steam than with the link motion, as there is less friction set up.

This engine has the regulator and cock arrangement for controlling steam to the cylinders similar to that in the previous locomotive (see pages 43 and 212; also Sheets No. 1 and No. 11, B).

The brake (see elevation, Sheet No. 12). Wooden blocks are screwed or pinned to hangers; these are operated on by the horizontal rod, levers, and screwed rod in the brake pillar, which is fixed vertically upon the side of the foot-plate on the inner side of the water-tank. The rods on each side are single, and pass outside the driving-wheels; the hangers are made of stiff sheet brass, a flat piece being left projecting from one side (see Sheet No. 12). This piece, when bent at right angles to the hanger with pliers or by hammering, forms the attachment for the brake-blocks. The hangers are soldered on each side to the ends of two stiff wires, which pass through them and revolve in bearings, or rather in holes, cut opposite each other in the frames. Collars are soldered on the wires bearing against the frames, which keep the hangers from shifting their position (this method of attaching the hangers to the frames is the easiest in a small locomotive). The bottom ends of the hangers are pivoted by means of fine copper-wire rivets to sheet-brass horizontal rods. The brake weigh-shaft (of wire) is carried in sheet-brass hang-down bearings, soldered at one side inside the frame-plates. The lever on the brake weigh-shaft operates the horizontal rods, and is itself moved by the lever and a pivoted nut, as already described for the tender and traction engine. The brake pillar may be a casting. There is a globular expansion of the screw at its upper end, which is embraced between hemispherical recesses, one half in the top of the pillar, the other half in its cap, and in this way the brake-screw turns freely by the hand-wheel, but cannot lift, and so elevates or depresses the nut and the lever, according to whichever way turned. The above should turn out a good working locomotive, and weigh fully in working order $6\frac{3}{4}$ lbs.

Locomotive with single driving-wheels (see Sheet No. 13). This locomotive being without coupled wheels, side-rods, and bogie-truck, is the simplest form to make, and will run the easiest of the lot. The boiler is constructed in the same way as those already described. The safety-valve, instead of being mounted on the dome, is fixed on a seating formed of a casting, or of sheet brass soldered together and placed over the fire-box. This is given for a variation, but the valve may be set on the dome as before. The side frames are cut so as to appear bent downwards in front of the smoke-box, for the purpose of keeping the buffers at the same height from the rails, as in the other locomotives; but this is not necessary. There is no transverse plate passing across below the cylinders for a bogie-truck to be supported under, but the frames are cut suitable for taking six wheels. The leading-wheel is small, in order to keep clear of the cylinder and piston-rod when revolving. There is only one guide-bar, an under one, for the piston-rod, the arrangement of the side-plates leaves no room for an upper one. The driving-wheels are placed in front of the fire-box, just far enough to allow the eccentric-sheaves and pulleys to clear the latter. The splashers are outside only, and the outer angles above the frames are curved to clear the connecting-rod; the splashers are represented as having a sand-box, from which, in front of the driving-wheels, a pipe for sand is placed. It will be noticed that the distance between the driving- and trailing-wheels is out of proportion to what it would be in a real locomotive; this is done for the purpose of getting in a large fire-box, as with fire-boxes proportionate in every respect these small engines will not work, and we wish to describe here only what, when made, will give satisfaction as good working locomotives. The trailing-wheel is much larger than the leading-wheel; this is in order to raise the axle above the lamp, so that it will not interfere with its being pushed in and drawn out. The shape of the cab is slightly different from what we had before. Of course the cab may be omitted, and only a wind-guard put up instead. The springs in front are "imitation ones," similar to those on the tender (see page 223). The connecting-rod is solid at both ends; either link motion or loose eccentric reversing-gear can be adopted. For dimensions, see Sheet No. 13. The sides of the cab and roof are soldered to the saddle-plate, and fixed at the foot on the top of the over-hanging frames by small bolts and nuts passing through angles on the cab and frames. The lamp must have fourteen $\frac{3}{16}''$ wick-holders. The barrel of the boiler and outer fire-box is made of No. 24 sheet copper. Internal fire-box, fire-tube, and tube-plate of No. 21 sheet copper, and frames of No. 20 sheet brass. The regulator and cock arrangement for admitting steam to the cylinders is the same as in the other locomotives. A tender suitable for this engine may be made from directions given on page 221, and from the drawings on Sheet No. 11, C.

A larger and more powerful working locomotive can be made by doubling (so far as practicable) the dimensions over all of the locomotive on Sheet No. 13. The boiler is made the same as described on page 46, and exactly twice the size of working drawings on Sheet No. 1. It must be brazed at every joint, fitted with a $\frac{1}{2}''$ safety-valve, a small grate, and with the pump described on page 71. To get the boiler in between the frames and leave a little room for expansion, the latter should be set at a distance of $5\frac{1}{8}''$ apart, but the thickness of metal required in them to give rigidity, along with the axle-boxes, prevent us from gauging the rails at $5\frac{5}{8}''$, so we must gauge them at $6\frac{1}{2}''$, and make other slight alterations here and there. The frames must be made of iron $\frac{3}{16}''$ thick, cut to outline by drilling and filing, then attached by bolts and (cast) brass angles to the buffer-beam in front, and to the cross-stretcher behind. The cylinders are of iron, placed outside, $1''$ bore, $2''$ stroke; steam is admitted to them by a $\frac{3}{16}''$ cock and pipe; blast-pipes $\frac{1}{4}''$ bore. The wheels can be made of iron, $\frac{3}{8}''$ thick, and $\frac{7}{16}''$ broad on the tread; flange $\frac{3}{16}''$ nearly; axles, of iron, $\frac{5}{16}''$ thick. The distance between the outsides of the overhanging frames is $9''$.

A tender may be made by doubling the dimensions over all of that on Sheet No. 11, C, and altering where required, to make it suitable for this locomotive. The whole thing can be made by soldering together sheet brass, and fitting axle-boxes on the frames; the tank ought to be utilized to carry a supply of water to feed the boiler.

A locomotive and tender such as this will run very well out of doors upon rails, and pull a considerable load behind it. The engine can be fitted with reversing link motion or loose eccentric reversing-gear, as desired. All the turning required to make this locomotive can be done on the lathe, described on page 3, with the exception of the driving-wheels, which had best be turned by the brass finisher after they are cast.

A **six-wheel coupled tank locomotive** is given on Sheet No. 14, such as is used for working branch-line trains and for shunting purposes. More friction is set up in this form of engine than in the others, from the coupling-rods being attached to six wheels. By adopting a short stroke and small driving-wheels, this locomotive will work fairly well, with twelve $\frac{3}{16}''$ wicks burning in the fire-box. The frames are 14″ long, and are cut to suit six wheels of equal size, these being each $2\frac{1}{4}''$ diameter. The front wheels are much nearer the central pair than the trailers are; this, as was explained before, is done in order to get a large fire-box in between the axles. The coupling-rod is jointed just behind the driving-wheel crank, for the purpose of diminishing friction. The cylinders are set so as to be inclined at an angle to the driving-wheels. There is only one guide-bar for the piston-rod. The foot-plate is entirely covered in by the cab, there being a doorway on each side. The back saddle-plate is cut for two circular windows, the same as the front one. As it is absolutely necessary to get your hand into the regulator when running the engine, we have made the roof of the cab portable, so as to lift off from the sides of the cab and coal-bunker (see drawings of roof in Sheet No. 14, and dotted lines in the elevation). The hind saddle-plate, the roof, and the sides of the doorways simply rest upon the sides of the cab and coal-bunker, and can be lifted off when required. They are kept in position by strips of tin soldered on the inside which fit against the sides of the cab, which are permanently fixed to the foot-plate. In order to clear the axle of the leading-wheels, the (wire) eccentric-rods should be bent into a semi-circular shape (see drawing), and pass below the axle. The boiler barrel and outer fire-box is made of No. 24 sheet copper, the internal fire-box, fire-tube, and tube-plate of No. 21 sheet copper, and the frames of No. 20 sheet brass. The engine can be fitted up to run in one direction only, or with reversing-gear, which may be either of the link motion or loose eccentric variety.

We have taken up one or two different types of locomotive engines, for the purpose of giving the amateur a few selections to choose his "model" from that he intends constructing. The reason why we have given examples of tank engines is that these locomotives, not being burdened with an almost useless (at least to them) appendage, viz. the tender, have more power, and prove more satisfactory under steam, than those provided with a separate tender.

We must bear in mind that, however well and carefully made a small locomotive is, it will never pull upon rails such a big load behind it as one would expect, at least to be at all in proportion to the amount of work done by a real locomotive upon the railway. This is owing to the fact, that in these small boilers, with all the heating surface you can obtain, steam of over 30 lbs. per square inch, or two atmospheres, cannot be generated and kept up. So that if one of these locomotives runs well upon a level railway, both backwards and forwards, and can at the same time pull two of the coaches to be shortly described, it may be regarded as a first-rate working locomotive.

It will be noticed that all the engines described in this book, as well as the locomotives, have a funnel or tube passing vertically upwards from the tank of the spirit-lamp just behind the fire-box, and through this tube spirits of wine is poured in as required, to keep the flames brisk when the engine is at work. This plan gives good results, and so we bring it before the notice of our readers.

Directions for working the locomotives, which apply to stationary engines as well:—

1. Oil the motion well, especially the eccentric-spindles, pulleys, piston-rods, coupling-rods, and axles. If oil can be put inside the valve-casing to oil the slide-valve, it is best to do so. Use a small oil-can, or dip a wire in the oil, and apply to the different parts. The oil used by watchmakers is perhaps the best for these small engines.

2. Unscrew the safety-valve casing, or water-plug, and fill the boiler with cold, but preferably

warm water, till it rises just to the top of the inner fire-box (as seen through the plug-hole); screw the plug or valve on again.

3. Pour spirits of wine or methylated spirits through the funnel (that leads to the lamp) behind the foot-plate or coal-bunker, and fill the tank about half full. Be careful not to put in too much, as the spirit is apt to overflow and prove troublesome. Light the fire by a taper or match pushed through the fire-door aperture. Steam rises from cold water in about five to eight minutes.

4. When steam is up, as indicated by its escape from the safety-valve (which must be properly set by tightening up the spring as required), pull the regulator outwards from the boiler, to admit steam to warm the cylinders. When it issues freely as dry steam from the blast-pipes, close the regulator by pushing it in towards the boiler, and wait till the steam gets strong (say about 30 lbs. per square inch); then place the reversing-lever in the first notch of the quadrant, or, if there be no link motion but only loose eccentric pulleys, push the engine by

hand forwards for a short distance to throw the eccentrics into forward gear. Then open the regulator, and the engine will run forwards. To stop, close the regulator. To *rcverse,* shift the reversing-lever from the forward to the backward notch in the quadrant, or if there be no link motion but loose eccentrics, push the locomotive backwards a short distance to throw the eccentrics into backward gear, open the regulator, and the locomotive runs backwards. To *stop,* close the regulator. Stoke the fire, and keep it brisk during the time steam is up, by pouring in spirits to supply the wicks as required. One of these locomotives will run from fifteen to twenty minutes after steam is up, without any fear of the water getting short, and so melting the solder. The brazed fire-box allows of this, because though the crown is uncovered by water, there is no solder there to melt.

5. When the run is over, draw the fire, by pulling out the lamp from under the coal-bunker, or if there be a tender, this latter must first be uncoupled from the engine to get the lamp out; then blow out, or smother the flame with a cloth; it cannot be readily

put out while the lamp is in the fire-box. Next replace the lamp by sliding it along the groove under the coal-bunker or foot-plate, and couple on the tender, if there is one. The water can be left in the boiler, as copper does not rust, or it can be emptied out by unscrewing the water-plug, and turning the engine upside down while holding it in the hands.

6. Occasionally the fire-tube and funnel should be cleaned, by passing a small round brush through them; the sides and crown of fire-box should also be scraped sometimes to remove soot (from the lamp-flames), which tends to gather to a small extent upon these parts.

7. Place fresh wicks in the lamp-holders when required: these may be composed of either cotton or asbestos twine. There is an advantage in using the latter, for it does not burn away so rapidly as the former.

8. When steam is noticed to leak through the eccentric and piston-rod stuffing-boxes, the glands must be unscrewed, the stuffing-boxes packed with fresh tow, wound round the rods, and the glands

screwed up again against the packing, to compress it and render the stuffing-boxes steam-tight.

How to pack Locomotive Glands.

Re-packing the locomotive glands with fresh packing (which requires to be renewed now and again, if the engine is run often) is rather a difficult job, but may be done as follows. For the piston-rod, lay the locomotive upon its side on a table, place the piston on the back dead-centre, or with the rod fully drawn out, unscrew the gland from the stuffing-box, push it well forwards on the rod till it is close to the cross-head, to give you room to work: it is not necessary always to take out the old packing. Take a long piece of tow, about the thickness of a thread, grease this well by soaking it in melted tallow; holding one end of the tow in the left hand, with the right hand wind it lightly round the piston-rod, making two coils, or at most three. With a sharp pair of scissors cut the tow short, and with a fine knitting- or darning-needle push it well down into the stuffing-box, equally all round the rod, screw the gland firmly up against the packing, to tighten it, and then slacken back the gland a little to free the packing, oil the piston-rod, and the whole is complete. If the piston does not move easily, slacken the gland a little backwards with a pair of watchmaker's pliers. The slide-valve glands are packed in the same way; when doing them the locomotive must be set on end, with the smoke-box downwards, or turned upside down, resting on wooden blocks. It is very important that the slide-valve glands be properly packed, or much steam will be wasted by blowing through them. A blow through one of these glands affects the engine more than when the pistod-rod glands leak. As regards the piston, if it is properly fitted, and packed with tow soaked in tallow when the locomotive is made, it will keep tight for a long time, and the re-packing of it may practically be ignored.

Note.—The locomotives with good strong steam can draw two coaches, but if steam is allowed to fall and become weak, none of them are able to move themselves, or if at work, they will stop and cause disappointment, when the real cause of their stopping is no fault in construction, but merely a want of power in the steam to propel them along.

Castings for making locomotives, with boiler materials, can be purchased from the model-maker's, and these vary in price from 10*s.* to 100*s.* and upwards, according to their semblance to a real locomotive.

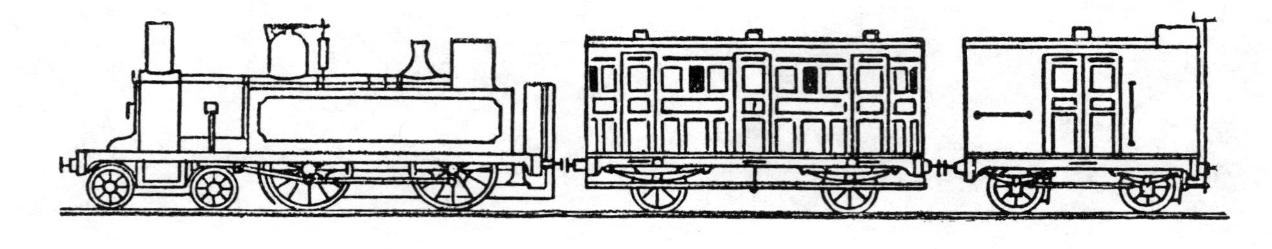

$\frac{3}{8}$ full size.

Fig. 55.

CHAPTER X.

CARRIAGES AND RAILWAYS.

Rolling stock (see Fig. 55 and Sheet No. 15). Here will be found working drawings of a correctly-designed three-compartment carriage, suitable for our railway. The carriage has only three compartments, which are quite sufficient for a model; having more simply means adding to the weight, and makes it more difficult to be pulled along. The frame is of wood, and consists of two longitudinal soles, or side timbers, extending from end to end, between the end timbers. These are narrower along the greater part of their length than at the ends, here they are $\frac{7}{8}''$ in depth. This is done to lower the buffer-beams, and keep the buffers on a level with those of the locomotive. The end timbers join the longitudinal ones at the ends, and are fastened to them by glue and small sprigs (for dimensions see drawings). The brass buffers are screwed into the end pieces, as well as draw-hooks, similar to those already described. Cross or diagonal pieces of wood need not be put in between the frames to stiffen them, as the end pieces do all that is required. The wheels are put inside the longitudinal timbers, which are placed at a distance of $3\frac{3}{4}''$ over all. The bearings are made of sheet brass, the axles revolve in holes drilled through them: they are cut to outline. Axle-boxes and imitation springs, similar to those on the tender, can be put on if desired; or the bearings may be carried up, and screwed simply to the sides of the longitudinal timbers; but as these are too thin to stand a wood screw, it is best to put a small screw through, two for each bearing, with a nut which can be tightened up on the inside and hold them firmly. The wheels are of brass, $1\frac{3}{4}''$ diameter, turned on the lathe, and firmly fastened to their axles, which pass through them before they enter the bearings, the same as in the tender. The axles are straight, and of the same thickness throughout. The wheels are kept from striking against the bearings when revolving by having projecting bosses, or narrow brass rings inserted between them and the bearings. These rings are slipped over the axles before putting them in place.

A foot-board is made of thin wood, $\frac{1}{8}''$ thick, and $\frac{5}{16}''$ broad, one on each side. Each is fastened to the frames thus: Cut a long strip of sheet brass (there must be three supports, one near each end, and one at the centre, each $\frac{3}{16}''$ broad), and bend into shape shown; all the strips must be cut to the same size. Drill a hole in each support through both projecting crank-ends, and fasten the foot-board to these, by one little bolt and nut for each. Fasten the sheet-brass strips transversely under the carriage frames by bolts, one on each side, and the foot-board is complete, after a stiff wire (bent to the same shape) has been soldered underneath the brass supports (below the foot-plate) as well as under the body of the carriage. This wire stiffens the supports, and prevents the foot-board from being bent in any way. A piece must be cut out of the foot-boards, opposite the axle-boxes, upon the inner sides.

The body of the carriage is best made of wood; tin-plate might do, but the first is preferable. The body is fastened to the frame by glue and sprigs (drilling holes in the wood first of all, as it might split unless this is done), which pass through the floor. The body must be attached to the frames before putting on the roof. The ends are quite flat, but arched at the top to suit the curvature of the roof. The carriage sides are both exactly the same, so that a description of one applies to the other. It is made of wood, $\frac{1}{8}''$ thick; the windows and doorways must be carefully marked out and measured, while the piece of wood lies upon a flat surface, then gouge out the windows, cut, and file to outline. The doorways had best be cut in the same way, as with such thin wood a saw might split it, but by gouging out the pieces, while the wood lies on a flat surface, this will not readily happen. The windows are six in number, $1'' \times \frac{1}{2}''$. The doorways are rather larger, and must be prolonged downwards till they meet the lower free ends. After this, nail the sides to the ends, and for this purpose use fine sprigs and glue. The two cross-pieces or divisions of the compartments are made the same as the ends, and fastened in their respective places. Fasten on the floor-piece, and nail a piece of wood right across upon each side of the doorways for seats, in all the compartments. For the roof, procure a piece of wood of the size given in drawing, plane it quite flat, and in order to bend it round to the correct curvature of the ends of the carriage, with a fine saw make a number of parallel longitudinal saw-cuts from end to end, going fully three-quarters through the wood—make all the saw-cuts on what is to form the inside of the roof—boil the piece of wood, and steam it well for two or three hours, and having planed the upper edges of the sides level, bend the wood over the ends of the carriage (which it will easily do while still warm), keeping the saw-cuts on the inside, and running in a longitudinal direction. The roof must now be nailed to the ends, sides, and divisions of the body. Drill holes through it, coat the under surface near the edge with glue, and nail with fine sprigs to the body of the carriage. Lay aside to dry, and afterwards plane the edges flush with the ends and sides of the carriage, or let them project over the latter a very little. Turn a piece of wood on the lathe, $1\frac{1}{2}''$ long, to a $\frac{1}{2}''$ diameter, cut this into three pieces, each $\frac{1}{2}''$ long; drill a hole through the centre of the roof of each compartment—this hole must be vertical—it need not go through, but should be nearly $\frac{1}{2}''$ in diameter; there are three holes in all. Glue one of the pieces of wood into each hole, this gives a finish to the carriage, and looks like one of the ordinary lamp-covers that are fixed on the top of a railway carriage.

The doors are six in number, made an easy fit in the doorways, each about $\frac{1}{8}''$ thick; a window must be cut out in the upper half of each. These doors may simply be made a tight fit in the doorways, or hinged by means of cloth and glue, attached on the outside, all down one side (the same as some of the small parcel-post boxes are hinged), or tiny hinges could be made of sheet brass and wire, these being fastened outside by small bolts and nuts to the carriage side of the doorway by one half, and by the other half to the door itself. Glue on wooden pegs for handles, or turn half-a-dozen brass handles out of a piece of wire, cut them apart, drill, and tap with a thread, pass one through each of the doors, and fasten inside with a small nut. Any kind of proper fastening for attaching or locking the doors would be troublesome to fit up; but if the doors are made a good fit in the doorways, they will remain tight when closed.

We must now begin glazing operations, to complete our carriage. If the amateur does not possess a diamond for cutting glass, let him give the sizes to a joiner or a glazier, who will cut them for him. Being so small, the pieces will be troublesome to cut, but can be done with a little patience. If the tiny panes be tightly fitted in the windows and doors, they will remain fixed without anything more being done to them. Be careful in fitting not to split the doors or the carriage sides when putting in the glass, and this need not happen if due care be exercised. The fitting of the window-panes may be omitted, but if done, adds to the appearance of the model. These windows cannot be raised up and down, as in an ordinary railway carriage, for this carriage is of too small a size to admit of window-frames being put in.

The coupling-hook is fastened to the end cross-bars, as in the locomotives, and attachment can be made from it by three small links of brass wire, each about $\frac{3}{8}''$ long, of oval shape, and made by bending a pin, cutting off the point and head, bringing together the ends, and soldering the junction where they come together; three links are sufficient. There is no coupling-screw, as this would be difficult to make with a right- and left-handed screw, such as it would require to have.

The carriage has panelling on the sides. This is done by glueing thin strips of wood on the outside, underneath the windows, and around doorways and doors, after the carriage is put together, and before it is painted. This renders the carriage complete, but is difficult to do, and may be omitted; in which case the carriage can be painted all of one colour, or the panelling can be represented by gold, white, or black lines, according to taste. The weight of this carriage when finished will be fully $1\frac{1}{2}$ lbs. By means of bent wires pushed through the sides and ends, and fixed by nuts inside, hand-rails can be fitted up in different places; even steps can be attached to the frames and ends of the carriage (these last are for the porters to get on to the roof to light the lamps, etc.). Steps are made by cutting out small pieces of sheet brass, $\frac{1}{2}'' \times \frac{3}{16}''$; bend each at the centre into a right angle. One half is fixed to the end of the carriage by bolts, the other half sticks out and forms a step.

Luggage van (see Sheet No. 16). It is constructed after the same manner as the carriage. The longitudinal and cross frames, the floor, body, and roof, are put together in the same way. Towards one end of the van, it will be noticed that the roof is raised for a distance of $1\frac{3}{8}''$, about $\frac{5}{16}''$ higher than the remaining portion. This elevation is formed of a solid piece of wood, cut concave below (to fit the roof) and convex on its upper surface; this piece is glued or nailed on to the roof. The doorway is wide, and the door is made in two halves, upon each side, and is hinged on both sides by small hinges, or by cloth and glue. Both doors open outwardly. Place a small window in one of the ends of the van, the same size, and put at the same height as those in the doors. The wheels are $1\frac{3}{4}''$ diameter, fixed to their axles and bearings, as in the previous vehicle. Foot-boards are attached in the same way, as well as axle-boxes and springs.

The brake, with its levers, rods, and blocks, is fitted up in the same way as that described for the tender. We have departed here in great measure from actual practice, and have carried up the brake-rod and hand-wheel or lever, and placed the latter on the top of the roof, as it is impossible to get the hand inside the doorway of the van to turn the brake off and on. The brake-pillar is a piece of sheet brass, screwed at the under surface to the floor of the van near one end, and the top is firmly screwed to the roof; the rod simply passes through two holes drilled in the frame, and is kept from rising and falling by a collar soldered on underneath the upper and under support, and above the frame by the boss of the hand-wheel, which is firmly screwed to the rod; the brake-pillar passes up inside the buffers, between them and the draw-hook. Panes of glass must be put in the doors and windows, and a piece of wood inserted in the roof to represent the lamp-cover. The wood of old cigar-boxes does very well to make the body of each carriage from, as this wood is about the correct thickness, is strong, and will stand nailing together with sprigs; if holes have been previously drilled for the sprigs to enter, the wood will not split.

A **truck** or **wagon** (see Sheet No. 16) is very easily constructed, and is the last vehicle that we intend taking up. The buffers are simply the projecting ends of the side timbers, the cross-pieces for the ends being between them. The floor is nailed on to the frames. The body can be made of separate pieces of wood joined together, or the sides and ends may be composed each of whole pieces, nailed together at the angles. The ends are curved upwards, and are highest at the middle. The wheels are $1\frac{1}{2}''$ diameter, and revolve in sheet-brass bearings, which are fixed by angles and small bolts to the under surface of the side timbers. There are no springs or axle-boxes.

Any other size of wheels, slightly larger or smaller than those given, will do for the carriages to run upon. If slight alterations are made in the bearings to correspond to the size of wheel adopted, and yet not raise the roofs any higher than the sizes here given, it would make the carriages seem rather out of proportion.

A drawing of the train will be found in Fig. 55, which is composed of a bogie-tank locomotive, a passenger coach, and a guard's van.

The **railway.** Two different kinds of railways suitable for the locomotives to run upon will be described; the gauge of each is $2\frac{7}{8}''$. The first form is easily and cheaply constructed, being made of iron wire, soldered to strips of tin, fixed upon cross-sleepers, nailed to a board. The second form is a correct model, having cast-iron rails, secured by wooden wedges in brass chairs, spiked to sleepers, as on the railway.

To make the *first*, procure iron wire of fully $\frac{1}{16}''$ thick (brass wire is not so good, as it produces more friction when the wheels are running over it), cut the wire into lengths, each about 40″ long, carefully straighten it by hammering with a light hammer over a flat piece of iron, or upon the vice; cut strips of tin, $4\frac{3}{8}$″ long and 1″ broad, nail these with a sprig near their ends to wooden cross-sleepers of about the same size, and $\frac{1}{4}$″ thick; fix these sleepers at a distance of 3″ apart from each other, in order to render the wire rails perfectly rigid. This allows about twelve sleepers to each section of line. Solder the wire down to the sleepers on the outside, upon one side, and at a distance of $2\frac{7}{8}$″ solder down the wire on the other side. The solder, it will be seen, takes the place of chairs, and holds the rails in position. The length of line, viz. 40″, is now complete. Any number of these sections can be made in the same way, and then the sleepers with the rails attached must be nailed down to a board, or a piece of flooring plank; the ends of one section of rail coming close up to those of the next section, so as to render the rails of the various sections all continuous with each other. Raise the board along with the rails upon wooden supports, fixed at a height of say about 40″ from the ground, and be careful that the line is fixed up level. Ascertain this by means of a spirit-level, after which the line is complete. It may be permanently laid down in any outhouse or workshop (as these locomotives do not steam well outside, unless the day is very calm), and is ready for use at any time. If this line is properly made, it will give satisfaction, and there is little fear of the vehicles leaving the rails, if the flanges are good, and the gauge correct; but care must be taken that the flanges of the wheels (when running) keep clear of the tin sleepers, or otherwise considerable friction will result, and may cause disappointment, as the engine fails to start, or stops very soon after being set in motion. A curved line produces more friction than a straight one. A siding, points, and a crossing may be easily made, by following the directions given under the second form of railway. The pointed end of the wire switches must be filed down very thin, where they come in contact with the stock rails. The two tongues of the points, in this case, should be soldered down to a strip of tin, which passes under the main-line rails, and rests upon a wooden sleeper, but is not attached to it. To the end of this tin plate, on one side, a wire is fastened which pivots to a lever in the ordinary way. Pulling or pushing this lever causes the points to be set for the siding or main line, as desired. The tin plate, with the points attached, will not move freely on the wooden sleeper below it, unless the plates attached to the second and third sleepers from the points slide also upon their respective sleepers. In this way the heel of the points can be soldered to the cross-plate, and no pivots are required. The crossing is made in the same way as that to be described further on, there being both wing and guard-rails made of wire. The rails must be bent away from the main line, and fixed to the sleepers with solder. Gaps must also be made, to allow the flanges of the wheels to pass.

The second form of railway (see Figs. 56 and 57—which are full size—and Sheet No. 17). A pattern for a rail and chair must first be made in wood (see pages 305 and 306), and castings obtained from them. Get the rails cast in malleable iron, as these castings are soft and easily manipulated. Each rail can be bent into a circular shape, or

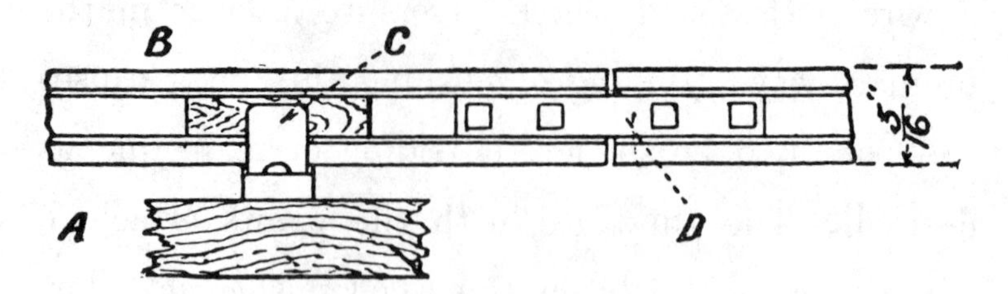

FIG. 56.—RAILWAY.

A, wooden sleeper ; B, rail ; C, chair with wooden wedge ; D, fish-plate.

hammered, without fear of it breaking, as an ordinary cast-iron rail would do. The rails, after being straightened by hammering on a flat surface, should

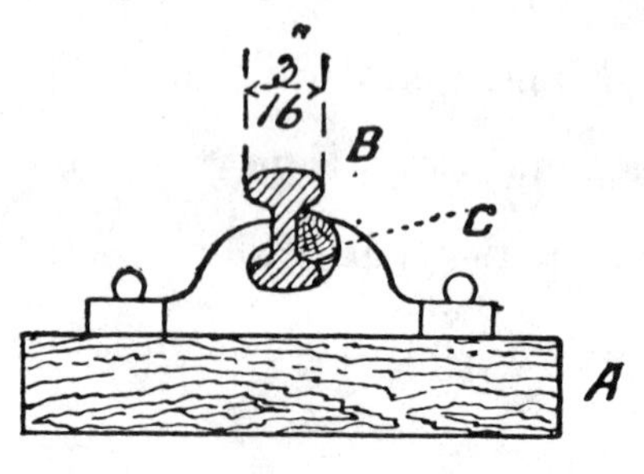

FIG. 57.

be filed over on the head, heel, and sides, then polished upon the upper and inner surfaces, by rubbing on the grindstone, and finished with coarse and fine emery paper and oil, then they will become as smooth as glass, and reduce friction to a mimimum, when the wheels pass over them. Get all the chairs cast in brass, file them flat on their under surface, and each must be drilled near the ends for two bolts, to spike them to the sleepers. To save expense, four chairs, or even three, might do for each rail. The sleepers should be $4\frac{3}{4}''$ long, $1\frac{1}{2}''$ broad, and about $\frac{1}{4}''$ thick, and these, instead of being embedded in ballast, should be nailed to a board, raised about 40″ from the ground, as before. Care must be taken that the inner jaws of the chairs do not project so far up against the web of the rail as to rub on the flanges of the wheels when passing over them. When fixing the chairs and the rails upon the sleepers, a measuring instrument or gauge should be used to see that the rails are set at the proper distance apart. This can be made out of sheet brass, having a piece cut out of it at each end, and leaving a prong, or wider piece between the rails. The ends of the gauge lie on the upper surface of the rails, and the distance between the outside edges of the upper part of the prongs is the gauge of the railway. A rigid connection of the rails endways can be made by fixing the ends of each rail into a joint chair, half of the chair going to one rail, and half to the other; or if we want a correct model, by joining the ends of the rails by fish-plates (see drawings), made of stiff sheet brass, each $1'' \times \frac{3}{16}''$ nearly, one being placed on each side of the web of the rail. The upper and lower edges of the plates are filed at an angle, to fit accurately the sloping sides of the head and foot of the rail. Bolts (either two, as on the railway, or one) should pass through the rail and the two fish-plates, and draw the plates together. Then tighten them up against the rail with nuts. Of course this necessitates very accurate drilling of all the holes in the rails and fish-plates, so that the bolt-holes may correspond. The bolt-head is placed on the inside, and the nut outside. A distance of from fifteen to twenty feet would be a fair length to make the line, a less distance than this is rather too short for a good run.

A siding can easily be made with the malleable rails (as they bend easily), running off from the main line. If properly gauged, and the curve correctly set out, the model will work well, and it is interesting to see an engine take the points and run round the curve under steam, dragging a coach along with it. A curve with a radius of ten feet will allow any of the locomotives and carriages mentioned above to pass round it, but the outside rail must be slightly elevated above the inside rail, with a gradual rise from the main line to the sharpest part of the curve. Points are movable rails, by means of which the direction in which a coach is travelling is changed, and crossings are gaps in the rails through which the flanges of the wheels pass when a coach on one line crosses to another line. Two drawings of a pair of points are given: in Sheet No. 17 the points are standing right for the main line, and also for a line diverging to the left. The wheels of the coaches are kept on the rails by the flanges on the inner sides, so the wheels follow the guidance of the rails, and the path they travel in each case is shown by the shaded lines. The points are ordinary rails, which must be filed down very thin at one end, and pivoted by the other. The pivot end is called the heel of the points, and must be so placed as to allow room for the flanges to pass easily between it and the stock rail. When a pair of wheels has been guided on to the diverging line, and continues to travel on this line, the flanges of the right-hand wheels will have to cross the rails on the near side of the straight line. This is done by making a gap in the straight rail, and to allow the wheels on the straight line to cross the diverging line, a gap is required to be made in the diagonal rails. The intersection of the rails at the gaps constitutes a crossing (see plan, Sheet No. 17). In order that the flanges of the wheels may pass through the gaps, the wheels near the crossing, and the opposite wheels, are guided by wing rails and guard rails, fixed near to and opposite the gaps, and these, acting as check rails, prevent the wheels from diverging to the right or left, and so pass through the gaps in the wrong direction. The crossing can be made from the main line at any angle so long as the gaps can be properly protected with guard and wing rails. In making a siding, the ends of the points must be filed down very thin, the extreme end made so as to be kept lower than the top of the stock rail, and below the level of the flanges of the wheels, and the pointed end must fit very accurately against the stock rail. The two tongues of the points must be rigidly connected, so that they may move accurately together, and that the horizontal distance between them may be properly preserved. Make the connection between them by means of a piece of brass wire, having collars on the inside for the rails to rest against. The free ends of the wire which pass through the rails can be riveted firmly on the outside of the web of each rail. To move the points, one of the connecting-rods is prolonged on one side, or a separate rod is used, cranked beneath the rails, and rising, so that it may be fastened to the point rail, or centre of the connecting wire. This rod is extended horizontally, and pivoted to a lever, which itself pivots in a bearing fastened at a proper distance from the line, to be clear of the trains when passing. This lever must be moved, when it is desired to alter the position of the points, or it could be fixed by a spring, so as to keep the points set for the main line, and only require moving when it is desired to run to or from the siding. At the crossings, the rails require to be filed away, so as to meet at the correct angle with each other, and be firmly wedged down to the chairs as well. The siding can be prolonged for any convenient length.

A simpler plan of arranging the points, and one that will save some fitting, is to adopt the original mode of providing for trains crossing from one line to another, and which is still used by contractors on their temporary railways. In this method (see Sheet No. 17) the four ordinary rails of the two diverging lines are brought as near together as space required for the flanges of the wheels to pass permits; and the two rails of the single line are left of the ordinary size, and are pivoted, so that their free ends can be placed opposite the ends of either pair of the diverging lines as required. The movement of these can be regulated by a lever as before, and the movable rails are kept fast together by a bar that passes between them. If the railway is made a straight one, it will give the best results, as a circular line always causes considerable friction, and prevents the trains from running so easily upon it. The same remark applies to wooden railways, whether circular or straight. Another plan of fixing the rails to the sleepers, which does quite as well, and saves getting brass chairs cast, is to set the rails direct upon cross-sleepers, nailed to a board at a distance each of 3″ apart, omit chairs, and use what practically amounts to chairs. This consists in forming each chair of two separate pieces of hard wood, each about $\frac{3}{4}''$ long and $\frac{1}{4}'$ thick (preferably from the wood of a cigar-box). Nail one piece outside the rail on a sleeper, and the other inside the rail opposite to it. The edge of each piece of wood should be made convex towards the rail, so as to pass inside the groove, and grip the web tightly upon each side when nailed down, and so hold the rail firmly—only the wood must be cut away on the inside, so as not to rub against the flanges of the wheels when passing over the wooden chairs. The rails may be attached to every second sleeper, which is quite sufficient, whether fish-plates are used or not. Make sure that the rails are properly gauged before nailing (with $\frac{1}{2}''$ sprigs) the two pieces of the chair down to the sleepers, or there is trouble afterwards to get the rails back to the proper gauge. Our reason for adopting cross-sleepers under the rails, and not fastening the latter direct to the wooden board, is in order to allow access of plenty of air to the locomotive fire-boxes, to support combustion.

If brass chairs are adopted to fix the rails (when laying them) wooden wedges must be cut, each about $\frac{3}{4}''$ long, and driven firmly in between the rail and chairs on the outside. It is best, and perhaps easiest, in order to gauge the rails properly, to wedge them to the sleepers first, and then, when fixed, bolt the chairs and the contained rail to the sleepers; lay down one rail first, and then lay down the opposite rail at a fixed distance from it, using the gauge to get the correct distance. It will be seen that a few sleepers are cast of a different form for the points and crossings, and patterns for these must be made specially.

If the line is laid down permanently in a workshop or outhouse, where there is the slightest dampness, the surfaces of the rails are apt to become tarnished over, or it may be coated with rust, which spoils their efficiency, and in order to keep them in good working order their upper surfaces should now and again be rubbed over with fine emery cloth, and smeared with a little oil.

Railway signal (see Sheet No. 18). We adopt the ordinary semaphore signal. The post is made of wood, dove-tailed, or otherwise fixed in the wooden board upon which the line is carried, and set near the junction (the correct distance is out of the question, as the line is too short to admit of this). The upper end of the post has a longitudinal slot cut out, through which the sheet-brass arm passes, after being pivoted inside. When the arm is fixed horizontally, this indicates danger, and when it falls nearly vertical it shows the line is clear. A small crank is soldered to one end of the wire that passes through the upper part of the signal-post; this wire has the signal-arm fixed upon it, and oscillates, in response to the movements of the crank, in a horizontal hole drilled through near the top of the post. The arm can be fixed inside the slot, or outside upon one side of the post; it is all the same, as both forms are used on the railway: the last is perhaps the easiest to fix up. A sheet-brass rod, or a wire, connects the crank to one arm of a weighted lever, pivoted near the foot of the post, while the extremity of the short end of this lever is attached to a small chain, which passes under a pulley-wheel, and along the side of the line, to be connected with a hand-lever (working in a racket) at any distance chosen from the post.

Make the arm of sheet brass; if placed outside the post, no slot is required. Drill a horizontal hole through the top of the post, attach the arm by solder to a wire, pass one end of this wire through the post, leaving the arm outside, put a nut on the end of the wire to prevent its being drawn out, and solder on a brass crank upon the other end for making attachment by to the levers. A nut fixed on the end of the crank-pin will prevent the long rod, going to the weighted lever, from coming off. If the arm is put inside the post, there is more difficulty in getting it fixed inside the slot. The best way to accomplish this is to make a saw-cut, from one side, right into the slot, to meet the upper and lower borders; take the piece of wood out, drill a horizontal hole, from each side, into the slot, put through the wire as before, solder the arm to it, and fit a piece of wood into the slot on the opposite side, and drill a hole through it for the arm to pivot in, then fix on an outside crank. After painting, and attachment has been made to the levers, the whole is complete.

A small chain, and any size of pulley, will do. The chain (it will be seen), when in the racket, passes underneath and upwards for a short distance behind the lever, and is attached to a snug on the back or farthest-away side of the lever from the signal. Both the racket and the lever may be made of brass, and set at any convenient distance from the signal.

A suitable railway for the large locomotive described on page 239 can be made in much the same way as the above, but with heavier rails, chairs, and sleepers. The latter will require to be each about 10″ long, and the former (if the line is permanently laid down out of doors in a garden) had best be made of brass; because small iron rails, if exposed to weather, will soon get rusted over and rendered useless. The brass rails will no doubt become tarnished, but this can soon be put right by fastening a piece of sandpaper to a piece of wood, and going over them now and again. The line could be raised on supports, but perhaps had best be fixed on the ground. Gauge of line is 6½″.

CHAPTER XI.

COMPOUND MARINE ENGINES.

Compound non-condensing marine engine. This engine differs from all those hitherto described, in being of the compound type. By this we mean an engine in which the steam, after having done a certain amount of work in a high-pressure cylinder, exhausts from it into a larger low-pressure cylinder, where it does further work, and from this last, in our engine, it exhausts direct into the atmosphere. In a real engine of the condensing type, the steam, after passing out of the low-pressure cylinder, goes into the condenser, where it is condensed into water. We have entirely omitted the condenser, with its circulating and air-pumps, from this engine, as being too complicated, and possibly might not work well when made on such a small scale. This engine is of the cross cylinder type (see Sheet No. 19). The cylinders are placed side by side, and each connects with its own crank. The two cranks are set at right angles with each other, and this being so, when the steam is discharged from the high-pressure cylinder, at the end of the stroke, it exhausts into the receiver B, which surrounds the high-pressure cylinder, then it passes by the pipe C to the low-pressure cylinder, and is admitted to it by the slide-valve. From the low-pressure cylinder, as mentioned above, it passes out into the atmosphere through the exhaust-pipe D. It must be understood that the letters H.P.C. stand for high-pressure cylinder, and L.P.C. for low-pressure cylinder, and these letters will now be used when referring to these cylinders.

Cylinders. The H.P.C. is $\frac{3}{4}''$ bore, and the L.P.C. $1\frac{1}{4}''$ bore, each with a stroke of $1''$. Both are made of brass; the flanges of the H.P.C. are of the same diameter and thickness as those of the L.P.C., so that the top covers of both cylinders can be cast from the same pattern. By referring to Sheet No. 19, it will be noticed that a considerable space (B) exists round the body of the H.P.C. and the edges of the flanges; this space is to be completely closed in, by means of a circular piece of sheet brass, extending above downwards from the edge of one flange to the edge of the other, and laterally from the steam-chest on one side (including the exhaust-pipe in its interior) passing round the cylinder, and is fastened to the opposite side of the steam-chest. This metal sheet is to be fastened to both cylinder flanges (all round) by means of small cheese-headed screws, screwed here and there into the flanges. And besides this, the edges of the flanges, the steam-chest, and the inside of the metal sheet (having been previously tinned) are to be soldered together above and below, and also to the sides of the steam-chest, so as to be made perfectly steam-tight. This space forms a jacket surrounding the H.P.C., and is the receiver, and here the exhaust steam, after coming out of the H.P.C., waits till the L.P.C. is ready to receive it. There is no outlet for the steam, except through the pipe C that is soldered into the receiver opposite the steam-chest, and which communicates directly with the L.P.C. steam-chest. A cock should be screwed into the receiver, to let out any condensed water. These cylinders are rather small to have blow-off cocks, but if a lubricator, with a cock, is screwed through the top cover of each, this will do both for oil and for blowing out any water that may condense inside. These lubricators had best be purchased ready-made. Steam coming from the boiler is admitted to the H.P.C. by a cock, or wheel-valve, screwed into the valve-casing. Both cylinders are bolted by four bolts to the front covers. When fitting, centre the cover on the upper surface, punch a hole here, to correspond with the centre of the cylinder bore when bolted to it, drill this hole vertically downwards for the piston-rod to pass through, now turn the standards upside down, and enlarge this hole for about half its depth from the under surface, and fit a stuffing-box gland to hold the packing.

Both standards are fitted up in the same way, and must be filed smooth, and bolted, by means of the flanges (cast on their free ends), to the bed-plate. The piston, covers, slide-valves, and cases, must be fitted up as already described. The pistons are grooved, and packed with tow; there are no piston-rings. The bearings for the crank-shaft are three in number, made of brass, and bolted to the bed-plate.

The crank-shaft can be purchased from one of the model-makers, and finished on the lathe; it should be $\frac{1}{4}''$ thick, and $\frac{3}{16}''$ at the journals. A built-up crank-shaft can be easily made by filing up two blocks of brass to the sizes $1'' \times \frac{1}{2}'' \times \frac{13}{16}''$ (see drawings, Sheet No. 19), drilling a $\frac{3}{16}''$ hole transversely through both, near one end, tapping with a thread for a $\frac{3}{16}''$ screw, and screwing in the separate pieces of the shaft. Before doing all this, slot out a longitudinal groove in each block (see drawings), so as to leave a thick piece of metal on each side, which forms the webs of the crank. The thick piece at one end is to form the crank-pin. Measure on the outside of each web a distance of $\frac{1}{2}''$ from the crank-shaft centre, mark and punch a hole here: this forms the centre for the crank-pin. The crank-pin end of each block must be rounded and made narrower than the other end, then file out the crank-pin, and see that it is made to coincide with centres already marked. If the pins be filed circular, the engine will work all right, without the necessity of their being turned on the lathe. The shoulders of each piece of the axle may be soldered to the webs, as well as screwed in, and this will make the axle quite strong enough. In this last case, where we employ a built-up crank, there will be no difficulty experienced in getting the eccentric-pulleys of the L.P.C. slipped on to the shaft, as this can be done before the parts are permanently fastened together. When the shaft is a forging, it will be impossible to get the eccentric-pulleys over it, unless they are each put on in two halves. A rather troublesome undertaking with such small eccentrics; but it may be done as follows (see drawing). This method applies to both eccentric-pulleys. After finishing the pulley on the lathe, and drilling the centre of eccentricity, make a file mark across the pulley on both sides, in such a direction that the hole for the crank-shaft is halved in two by the file mark. From the bottom of the groove around the circumference of the pulley, drill a vertical hole that will pass right through the true centre of the pulley, and be at right angles to the file mark; drill this hole a good way deeper down than the centre of the pulley. With a small saw cut the eccentric in two halves along the file mark, widen the hole in one half of the pulley, for about half its depth—the head of the bolt, after being countersunk in the eccentric groove, rests in this wide part—tap the lower half of pulley with a small screw-thread, make a screw with a long head, so that the neck will rest against the narrow part of the hole in the upper half, and having slipped each half of pulley over the shaft, screw them together. Then file the head of the screw flush with the bottom of the eccentric groove, and the whole is complete. The eccentric-strap and screw will keep the two halves in line. These pulleys, being in two halves, had best be soldered to the crank-shaft, after finding their correct positions (see Chap. IV.), as a key might tend, when driven in firmly, to break the screw, and separate the two halves from each other.

The valve motion requires little to be said about it, as all is rendered plain from the drawings. The slot-links and lifting-links can be made out of soft iron by drilling, filing, and cutting to outline. A bearing is screwed to each standard, to support the weigh-bar. There are two double lifting-links to each slot-link, but a single one would be simpler to fit up, and do well enough. By referring to page 139, you will find how to fit up the link motion. It will be a good plan, after finishing, to lag the cylinders with small strips of wood, $\frac{7}{16}''$ broad, the lagging extending from the valve-casing of the H.P.C., over the outside of the receiver, to the L.P.C., passing round it on to the opposite side of H.P.C. valve-casing, and from above downwards, passing from the edge of one flange to the lower edge of the other. Fit on a small plate of sheet brass, to close up the space between the cylinders, both above and below, attach these plates to the upper and under surfaces of the lagging by screws, and fill the interior of this space with plenty of horse-hair, as this is a bad conductor of heat. The lagging is kept in position by two brass bands passing round it, one near the top, and the other near the bottom of each cylinder; these can be fastened by pins or solder at their ends. The lagging will tend to keep the cylinders hot, and give a neat appearance to the engine when finished.

The connecting-rods are of iron, with the big ends of brass (see page 97); they are forked at the small ends, so as to embrace the cross-head pin. The forks may be cut out of the solid end, which is left thick. The cross-heads are of brass, or iron, screwed to the piston-rods, and these slide simply between the flat sides or guide-surfaces of the standards. A disc of iron, or brass, is fixed by a set-screw at one end of the crank-shaft, having four holes drilled through it near the circumference, for bolts to pass through, and join it to a similar disc on the end of the propeller-shaft. If the disc is made heavy, it will act the part of a balance wheel or fly-wheel, and save putting one on. The screw propeller can be purchased ready made, and mounted on a turned shaft, the same size as the crank-shaft; the propeller may be keyed on, or fixed by means of a set-screw. The length of this shaft will depend on the size of the boat in which the engines are to be placed. This shaft is coupled to the crank-shaft by means of bolts, passing through both discs, as mentioned above. The propeller-shaft must pass through a long stuffing-box, which goes through the stern of the boat, so that no water can get access along the shaft to the engines.

The dimensions of this engine will be got by referring to the working drawings, but the amateur will require to make special patterns to obtain castings. The patterns must all be made a little larger than the finished sizes, to allow of filing, and turning on the lathe. The L.P.C. steam-ports are $\frac{3}{32}'' \times \frac{9}{16}''$, exhaust-port $\frac{1}{8}'' \times \frac{9}{16}''$, travel of slide-valve $\frac{1}{4}''$, slide-valve $\frac{5}{8}'' \times \frac{15}{32}''$. Sides of valve-casing are $\frac{1}{4}''$ thick, to allow of bolt-holes, top and bottom borders, each $\frac{3}{16}''$ thick. The H.P.C. steam-ports are $\frac{3}{32}'' \times \frac{7}{16}''$, exhaust-port $\frac{1}{8}'' \times \frac{7}{16}''$, travel of slide-valve $\frac{1}{4}''$, slide-valve $\frac{1}{2}'' \times \frac{15}{32}''$. The top cylinder covers are both cast from the same pattern, $2\frac{1}{8}''$ diameter; these, as well as the cylinders, are made of brass or iron. The standards have guide-surfaces between the legs and the upper part; these can be either of brass or iron. The bed-plate is a plate of iron, or brass, $\frac{7}{16}''$ thick, having holes cast or slotted out for the crank and connecting-rod end to pass round when revolving. The two cylinders stand on their respective supports, quite independent of each other, being kept in place by bolts and nuts passing through the covers. They are joined together by the steam-pipe (with or without flanges) soldered on one side to the H.P.C. receiver, and to the valve-case of the L.P.C. on the other. For further information about attaching the cylinders to their front covers, boring and fitting the stuffing-boxes so as to get them in the central line of the cylinder bore, see pages 100 and 165.

Action.—From the drawings, it will be apparent that the steam coming from the boiler is admitted by a cock, or stop-valve, through the pipe A to the valve-casing of the H.P.C.; here it is controlled by the slide-valve, and after doing work in the H.P.C., it enters through the exhaust-pipe into the receiver B; from thence it passes to the L.P.C. by the pipe C, and after doing further work there, it exhausts direct into the atmosphere, through the exhaust-pipe D, fixed in the valve-casing of the L.P.C.

The **compound tandem engine** (see Sheet No. 20) is the form of engine used in all the White Star liners. This engine resembles the last one, but is of simpler construction, and can nearly be made from the same patterns. The cylinders are arranged in tandem fashion, so that the H.P.C. is placed on the top of the L.P.C., instead of side by side, and consequently both work upon the same crank-pin; the piston-rod and valve-spindle are common to both cylinders. This engine is much simpler to fit up than the last one, there being fewer parts, and only one crank on the axle; but the fitting up must be very carefully done, to get the cylinders and steam-chests in line, to ensure success when working. There is no receiver surrounding the H.P.C., as both cylinders take steam at the same time, and it is led direct from the H.P.C. exhaust-pipe to the valve-chest of the L.P.C. Patterns for a separate top and bottom cover for the H.P.C. must be made, the under one only having a stuffing-box and gland. The top and bottom covers, with the standards of the L.P.C., are cast from the same pattern as was used for the last engine. Both L.P.C. covers contain a stuffing-box and gland, as the piston-rod passes through the top cover, on its way to join the H.P.C. piston. The H.P.C. must be raised a good way above the L.P.C., in order to allow of room to unscrew the glands, and pack the piston-rods. This is accomplished by means of a special casting, a "distance piece," which sits upon the L.P.C. cover and supports the H.P.C., which rests upon it, and is attached by bolts to both cylinders. For dimensions, see Sheet No. 20, and how to construct patterns, see page 308. Get the distance piece cast in brass, with a bevelled hole through its interior. This casting must be mounted on a temporary wooden shaft, turned and faced in the lathe. The longitudinal slots in the sides can be drilled, and filed out to shape. The cored passage can be filed smooth with a round file, after which bolt-holes can be drilled through the flanges. This is better than casting the covers and distance-piece in one; for, being so small, it would be next to an impossibility to drill holes for the piston-rod, and fit the glands correctly. One standard only is required, having the bottom cover of the L.P.C. cast upon it. The piston of the H.P.C. can be screwed on to the top of the rod, and a nut put on to secure it. The best way to attach the piston of the L.P.C. is to make it a tight fit upon the rod, drive it on, then braze, or use hard solder, and it will be quite firm. The valve-casing of the L.P.C. has two stuffing-boxes and glands; the valve-spindle passes through a stuffing-box above and below, the spindle being common to both slide-valves. H.P.C. steam-ports are $\frac{3}{32}'' \times \frac{7}{16}''$, exhaust-port $\frac{1}{4}''$, travel of valve $\frac{1}{8}'' \times \frac{7}{16}''$, size of slide-valve $\frac{1}{2}'' \times \frac{15}{32}''$. L.P.C. steam-ports are $\frac{3}{32}'' \times \frac{9}{16}''$, exhaust-port $\frac{1}{8}'' \times \frac{9}{16}''$, travel of valve $\frac{1}{4}''$, size of slide-valve $\frac{5}{8}'' \times \frac{15}{32}''$.

The bed-plate is similar to the last, but contains only one slot. There are two bearings, of brass, one on each side of the standard. To fit these up, see page 165. There is no need of a weigh-bar to raise and lower the link by, but two lifting-links are pivoted direct to the reversing-lever working in the quadrant. There are two eccentrics, one for forward, and one for backward motion, and these control both slide-valves upon the one spindle. The eccentric-pulleys are put on the shaft outside the bearings, one end of the shaft carries a disc, as before. This engine could be fitted up without a link motion, and this will simplify matters very much.

Action.—The steam enters the H.P.C. from the boiler by the pipe A, and after exhausting from this cylinder, passes through the pipe B to the L.P.C., and after doing further work there, it exhausts into the atmosphere through the pipe C. No receiver is required, as both cylinders are ready to receive steam at the same time, and consequently it passes direct from the H.P.C. to the L.P.C. Lagging may be put round both cylinders, or they may be left as they are represented in the working drawings, without any.

CHAPTER XII.

PROPORTIONS OF ENGINES, ETC., PATTERN-MAKING, CASTINGS, ETC.

Proportions of Engines and Boilers.

As a rule, if the boiler for any engine is made five times as large as the cylinder, it will be of sufficient size to steam well.

To find the thickness of metal suitable for a boiler to stand a pressure of 15 to 16 lbs. per square inch (more than sufficient to run any stationary engine), divide the diameter of the boiler in inches by 100, this gives in decimals the thickness of metal suitable for the above pressures.

A good and safe method to find out what pressure a model boiler will stand, is to fill it quite full of water, load the safety-valve twice as much as you intend to use, and light the fire; as the water heats, it expands, and either will lift the safety-valve or the boiler will quietly separate without an explosion: this is called the "cold method" of testing a boiler. A vertical brazed copper boiler, 16″ × 8″ diameter, fire-box 7″ × 8″, with a 1½″ central flue, if heated with a gas-burner (too small for coal or coke), will drive an engine with a cylinder 1″ bore, 2″ stroke, fly-wheel 5½″, cut off at three-quarter stroke.

In small boilers, to get sufficient draught of air through the furnace to make the fire burn, the horizontal flue-tube should always be at least 1¼″ diameter (inside), and the funnel should never be less than from ⅝″ to ¾″ diameter. These sizes are suitable for burning methylated spirits of wine, or very small blocks of wood steeped in paraffin oil or methylated spirits. Smaller tubes than this do not keep up steam, from shortness of draught. Charcoal and dry wood will burn well in a boiler with a flue of 1½″ diameter, though there may be a little difficulty in getting the fire properly started. Common coal and coke require a flue about 2″ diameter, and a fire-box at least 6″ square.

As considerable time and trouble has been spent on the working drawings, in order to get good proportioned engines, and models that will work well when made, we think that the sizes of boilers adopted, with the great amount of heating surface provided in each, and the quantity of wicks burning in the fire-boxes, will prove satisfactory.

Notes on Painting Engines.

The cylinders of stationary engines require painting, and these can be made of a green, or any other suitable colour, by using ordinary paint mixed with copal varnish. The covers and bed-plates ought to be polished bright all over, and not painted. The wooden stands may be stained, either a mahogany or rose-wood colour, according to taste, by using "Castle Brand's" staining fluid, and putting on three or four coats. The spokes and insides of fly-wheel rims may be painted.

The **locomotives.** Before painting, remove all

roughness from the outer surfaces with fine emery paper, and grease with naphtha. Afterwards begin painting the boiler and engine, using a coat of red lead as a basis. When this is dry, paint the boiler, frames, wheels, cab, cylinders, and tender a green colour, smoke-box and funnel black, inside of cab and tender vermilion, or other suitable colours. Give two or three coats of paint, then finish with black and white stripes. Always paint a boiler when it is warm (hot water should be poured inside before beginning to paint), as in this way paint adheres better, and is not so liable to blister and peel off, after steam has been got up once or twice. Mixing copal varnish with the paint before using it, saves trouble in laying on a coat of varnish after the painting is finished, and looks nearly as well. Of course the varnish need only be mixed with the third, or last coat of paint.

Pattern-making and Castings.

We should like to devote a long chapter to pattern-making, but want of sufficient practical experience prevents our doing so; however, we shall give one or two hints upon this subject.

When going to make an engine, of whatever kind it may be, stationary, locomotive, or marine, prepare a drawing of the engine, make working drawings of the various parts, and mark off on these first the centre lines, in red ink, after which mark off from the centre line the dimensions of that part, so as to insure that it really is the centre line of that part. The drawings may be made full size, or to some convenient scale; in any case, always put in the distance, or dimension lines, as trouble will be saved in getting the dimensions afterwards from the drawing, by applying a rule, or a suitable scale. When castings are required, and the amateur has decided to make them for himself, patterns must be made in wood from these working drawings, and it is best to make the patterns always a shade larger in all their dimensions, so as to allow for the castings being worked on the lathe, and made to the requisite size. For small engines, shrinkage of the castings may be ignored. Pine, deal, or mahogany, with straight grain, are perhaps about the best kind of wood for pattern-making. Always taper the parts slightly which enter deepest into the sand, as this allows them to be more easily withdrawn. Avoid sharp internal edges upon patterns, and join the parts together with glue, or small sprigs. After the patterns are finished, and smoothed over, they ought to be varnished, or better still, brushed with blacklead, as then better moulds can be made from them. For the engines illustrated in this work few patterns will be required, except for the fly-wheel, cylinder, valve-case, bed-plate, and bearings, all the rest of the parts (including the piston) can be made out of brass rod, brass wire, and sheet brass, and these can be purchased square or round, of the requisite size. The patterns for wheels, either drivers or fly-wheels, when small, are best cut out of solid wood. The wheel is first turned as a disc, and with a flange; if a locomotive wheel, the boss is left wider than the rim; after turning, mark off the boss, the rim, and the spokes, lay the wheel on a hard piece of wood, and cut out the spaces between the spokes. Next reduce the spokes to the proper size and thickness, with a sharp knife and a file, then file the parts, smoothe, and finish with sand-paper, coat over with black-lead, and the pattern is complete. Never drill a hole for the axle in the pattern, but do it afterwards in the casting. Wheel patterns have a tendency to become a little oval in time; still, if made of a diameter larger than what the finished wheel is to be, it will not matter much though the pattern does get a little oval. Another method is: soon after making the pattern, before it has had time to become oval, to get a casting from it, and use this metal pattern for all future wheels, as it does not change in any way.

Note.—The small wheels should be cast in brass, or gun metal, which is much easier to work than iron. A good deal of trouble is saved in making the wheel-pattern, by having no more spokes than are absolutely necessary to render the wheel strong, and hold it together. Still, we do not advise fewer spokes than twelve in a locomotive driving-wheel, and perhaps six in a fly-wheel for a stationary engine.

The **cylinder** pattern (see Fig. 58) must always

be made a size larger in all its dimensions than the finished cylinder is to be, so that after the casting is turned on the lathe and completed, it may agree with the working drawings.

It is easiest in these small cylinders to make a

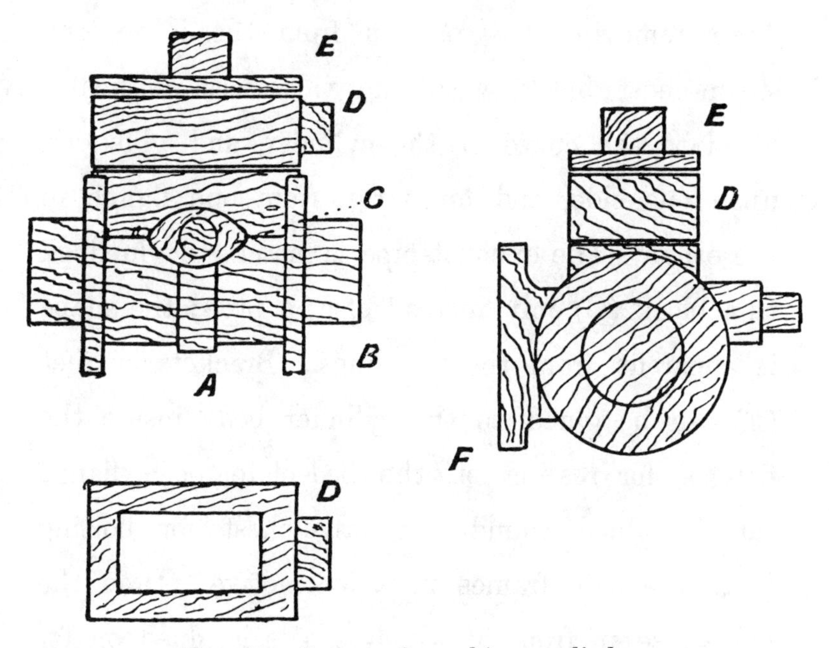

FIG. 58.—Patterns for making a cylinder.

separate pattern for the steam-chest, cast it quite distinct from the cylinder, and bolt it on afterwards. For the body of the cylinder, take a piece of hard wood (chuck it between the lathe centres), sufficiently long to include the body A and the cores B at both ends. Turn the flanges, and the body as well as the cores; these last should be left about $\frac{1}{16}''$ less in diameter than the finished bore is to be. The cores enable the moulder to cast a cored-out passage through the cylinder. After removing the pattern from the lathe, the steam-chest block, with the valve-face, is neatly fitted to the curve of the cylinder, and glued on upon one side, and must run from one flange to the other. The exhaust-pipe is made and glued on as a block (C) only, instead of as a pipe; the shape is apparent from the drawings. Brackets or feet (F) can be glued to the cylinder body inside the flanges, for resting on the bed-plate, or a flange can be glued round the valve-chest for bolting it to the side frames in a locomotive. D is the valve-case, in front of which a boss is glued on for the slide-valve stuffing-box. The cover E fits the outer side of the valve-case. The steam-pipe is simply a block, and not a pipe, glued upon the outside. Steam-ports are never drilled in patterns, but are made afterwards in the cylinder casting, in the way already mentioned. Separate patterns are made for the covers, which are simply discs. When making a locomotive, both cylinders should be cast from the same pattern.

A pattern for a **rail** is made by sawing and planing a piece of straight grain yellow pine or box-wood to the following dimensions: $18'' \times \frac{3}{4}'' \times \frac{1}{4}''$. Lay the wood on a bench, one end of it resting against a support, take a $\frac{3}{16}''$-groove plane (rest one side of the plane firmly against one side of the piece of wood, the opposite side of the wood being fixed against two nails driven into the bench to steady it), and cut out a longitudinal groove throughout the full length of the piece of wood to a depth of $\frac{1}{16}''$. In order to keep the plane from grooving the wood too deeply, it may be necessary partially to block up the groove in the plane by a piece of wood driven a tight fit in between the sides at the bottom of this groove, to prevent the chisel in the plane cutting the groove any deeper than $\frac{1}{16}''$. Reverse the piece of wood, and plane a corresponding groove (leaning the plane laterally against the same side of the wood as before, though this is turned upside down) exactly opposite to, and cut to the same depth as the first groove. The piece of wood has now somewhat the shape of a rail, with a head and a heel and a central web, which should be left about $\frac{1}{8}''$ thick. Take the pattern, and with a smoothing-plane remove all the superfluous wood equally on both sides above and below the web, so as to leave the head and heel each $\frac{1}{16}''$ thick and $\frac{1}{4}''$ broad. All the corners should be rounded off with a smooth file, and not left sharp; both head and heel must be made slightly convex on their outer surfaces; in short, the rail must be made a correct model. Never drill bolt-holes for fish-plate bolts in the patterns; these can be drilled afterwards in the castings, if required. After filing, polishing with sand-paper, and coating with blacklead, the pattern is ready for the foundry, and any number of rails can be cast from it, preferably in malleable iron, or Swedish steel. Sixteen rails will be required to lay a line 12 feet long. The drawings of the rail on Sheet No. 17 are full size.

A pattern of a chair, or support that lies between the rail and sleepers, is given in Fig. 59. Make it the size of the drawing, having a slot through which the heel and web of the rails pass. The inner jaw rises upwards and curves inwards, so as to rest against the inner side of the web of the rail, and a wooden wedge is driven in between the outer jaw and the rail, to hold it in position when the chair is fastened to the sleeper. The

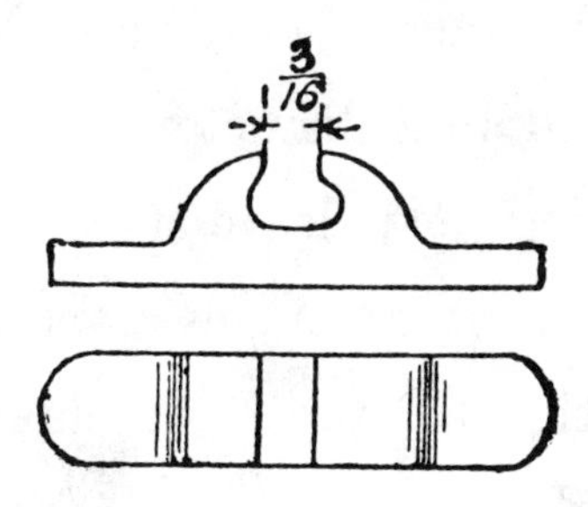

FIG. 59.—Elevation and Plan of Pattern for Chairs.

central slot must be made smaller than it is to be in the casting, and may be made to the correct size in the latter to suit the rail by filing. Get all the chairs cast in brass.

Both standards for the marine engine (page 284) are cast from the same pattern. The pattern is cut out of a wooden block to shape, and finished by filing and sand-paper. Cover it over with black-lead before sending it to the foundry; sometimes it is varnished instead. The flanges for bolting to the bed-plate by are made separately and attached by glue and sprigs. The bottom cylinder cover is fixed on at the top end. This is a separate pattern, turned to the proper size, and then fixed in position, with the centres truly in line. A special pattern for a "distance piece" (page 293) must be made according to drawings (see Sheet No. 20). The pattern is turned to a diameter rather larger than the dimensions given; leave a core at both ends projecting beyond the flanges, tapered slightly to correspond with the taper which the cored-out passage through the interior of the casting must have. The longitudinal slots had best be drilled out afterwards in the casting.

Tools required for pattern-making are a tenon saw, a small axe, a screw-driver, one or two sets of different-sized chisels and gouges, a small glue-pot, a pocket-knife, some few sheets of coarse and fine sand-paper, varnish, and blacklead, along with a set of wood-working tools for the lathe, comprising a set of chisels and gouges, which may be purchased in blocks.

Castings for making good working engines can be had from any of the numerous model-makers, and without recommending the castings of any particular maker, which we feel we are not in a position to do, as we have always made our own patterns, we will mention the names and addresses of one or two model-makers where good castings can be obtained—Mr. R. A. Lee, 203 Shaftesbury Avenue, London, W.C.; Messrs. Butler Bros., 135 Bentham Road, South Hackney, London, E.; Messrs. Lucas and Davies, 67 Farringdon Road, London, E.C.; Messrs. Stiffin and Co., 51 Roding Road, Homerton, London, E.; Mr. Bateman, 205 and 206 High Holborn, London, E.C. Copper tube, sheet, and copper rivets can also be had from Messrs. Stanton Bros., 73 Shoe Lane, London, E.C.

CHAPTER XIII.

HOT-AIR ENGINE, SMALL-POWER ENGINE, AND NOTES.

A hot-air engine. This engine is easily made, is a very safe one—there being no boiler which might be liable to explode—and when once started it will work away for hours without any attention, if a small gas-jet or spirit-lamp is burned in the furnace, and a proper circulation of cold water is kept constantly flowing over the heater to keep the parts cool. The engine is composed of the following parts (see Sheet No. 21). A is the working cylinder (which contains a trunk piston) with an inlet-port, which by means of a pipe communicates with the heating cylinder B. There is no cock on this pipe, but one could be fitted on. The under surface of B is heated by a spirit-lamp burning inside the fire-box C, and the top of B is cooled by a water-tight casing D, in and out of which a constant flow of cold water must pass. E is a compression piston, packed with asbestos twine; the plunger-rod has screwed upon it (a short way above this piston) a small piston F, which is packed with tow. This fits and works in a small cylinder, which passes through the centre of the water-tight casing above the heater. The upper end of this piston-rod, after passing through the small cylinder, is slotted out, and pivoted by a bolt to a connecting-rod, the cross-head of which engages with the small centre crank on the fly-wheel shaft. This small crank leads the crank of the working cylinder by an angle of about 80°. For dimensions, see the working drawings.

The engine can be made from No. 18 sheet brass, except that the bottom of the heater should be of No. 18 sheet copper. The furnace is circular in shape, brazed together, and to the bottom of the heater, having a fire-door at one side. And from the opposite side a funnel comes out at right angles, and joins by an elbow-joint to the upright funnel.

Rivet on a small hinged door, and also three feet, to raise the fire-box from the floor, and allow air to enter to support combustion. The lamp is made of tin, of a circular shape, having one small central wick, $\frac{1}{2}''$ diameter. There is a funnel to pour in methylated spirits by. The lamp sits inside the fire-box. All the joints of the funnel, the heater, and the water-casing should be brazed together, with the exception of the upper disc; this may be soldered on afterwards. The heating cylinder and water-tight casing should be in one piece; this forms a cylinder with a brazed longitudinal seam. A circular copper disc is fitted and then brazed inside near the middle, to divide this cylinder into two equal parts, the heater and the water-casing. Another disc should be brazed or soldered to one end, that which forms the top of the water-casing. The free end of the cylinder is afterwards to be brazed to the top of the fire-box, a disc being fixed between them. The centres of the upper and middle discs must be marked and drilled before fixing them in their places, and afterwards these holes can be widened to the proper sizes to allow the small cylinder G to be pushed through them both, the ends of which are then soldered to the discs (that to the bottom disc must be fixed before the fire-box is attached). A piston must be turned to fit this cylinder accurately; this piston is screwed on to a rod coming from the compression piston; the lower end of this rod has brazed upon it two copper discs, at a distance of about 1″ apart. The space between them is packed by rolling asbestos twine round the rod until it is filled with it, and the twine is flush with the edges of the piston. The two discs attached to the rod, with the packing, constitute the compression piston. The upper end of this piston-rod, after it passes through the small piston, is slotted out and pivoted to a connecting-rod, which engages with the centre crank. Having now finished all the parts, we take the compression piston in one hand, and the heating cylinder with water-casing in the other (the fire-box not yet being attached). From the lower end push the piston-rod of the compression piston through the central vertical cylinder, until the small piston enters it, when it will be found, if the correct distance is observed between the two pistons, that when the upper one enters the cylinder, the compression piston just enters the heater. The heater must now be brazed by its free end to the top of the fire-box, a copper disc being fixed between them.

Connection must be made by means of $\frac{5}{16}''$ pipes with the water-tank, and the casing D, in order to provide a constant circulation of cold water. The working or power cylinder is a brass casting, bored out on the lathe, the same way as a steam cylinder; it is open at the upper end, and has one port only at the foot. It is fixed vertically to and communicates by the port with the interior of the heater. A bottom cover can be cast on, or attached by studs. The pipe from the heater to the cylinder is $\frac{3}{16}''$ bore; this can be screwed into the cylinder, and soldered to the heater. The cylinder should be surrounded by a water-jacket made of sheet brass, communicating at the top with the cold-water cylinder over the heater, and underneath having a cock to let out water by. A trunk piston—*i.e.* a piston hollowed out in its interior (see working drawings), with a short piece of brass screwed to its centre, which is drilled and slotted out to pivot to a connecting-rod —is fitted in the working cylinder, and is attached direct to the connecting-rod, which engages with the crank on the end of the fly-wheel shaft. The crank-axle can be $\frac{1}{4}''$ diameter, turned to a little less at the journals. It can be purchased ready-made, of solid steel, with a $\frac{1}{2}''$ throw, in which case an end crank must be fixed on upon one side, suitable for a $1\frac{1}{2}''$ stroke, and set so that there is 80° degrees between it and the other crank. Or the crank might be a built-up one, with the webs of brass screwed and soldered together (see pages 105 and 285). The shaft revolves in bearings, supported on pillars fixed to the top of the circulating tank.

The water-tank is made of tin plate, or sheet brass, bent into a circular shape, and soldered together, with a flat bottom. The mouth is open, for pouring in a fresh supply of water now and again. Fix the tank as near the engine as possible. Connection is made from this tank to the circulating tank over the heater by two $\frac{5}{16}''$ bore brass pipes, and a better circulation of water is kept up by connecting the tube from the bottom of the engine tank with the top of the circulating tank, and *vice versa.*

The water is also allowed to circulate from the circulating tank round the power cylinder inside the water-jacket, from whence it can be let out by a cock; or if this cock is kept slightly open, water will continue to circulate. There is also a cock entering the circulating tank near the foot, which is used to draw off the water over the heater when this begins to warm, and allows fresh water to take its place. The fly-wheel may be either of brass or iron, and as so few castings are required for a hot-air engine, we think the amateur will be able to make all these from his own patterns.

Directions for working. Fill the tank quite full of cold water, pour spirits of wine into the spirit-tank, and light the fire; oil the motion, especially the working piston. In a few minutes, on giving the wheel a turn or two by hand, it will begin to revolve, and will continue doing so as long as a proper circulation of cold water is kept up, and the spirit kept burning in the fire-box.

The action seems to be explained as follows. The compression piston first compresses the cold air into the lower part of the heater when descending, where, on being heated, a greater increase of pressure occurs, corresponding to the increase of temperature, and this impels the working or power piston up to the end of its stroke. The compression piston having now moved upwards, transports the air alternately from one end of the cylinder to the other, and here it is cooled by the water-tank, and restored to atmospheric pressure, the pressure falling to a minimum, and then the power piston descends and completes the stroke, when compression again begins, ready for the second stroke, and so on.

Note.—If a leak occurs anywhere, this spoils the efficiency of the engine. If the water over the heater gets too hot, the engine stops, as the contained air can no longer be expanded and contracted as before.

This engine makes a powerful working model, but an engine half the size of the working drawings will work fairly well. Working hot-air engines can be purchased from Mr. Seal, 67 Carthew Road, Hammersmith, and vary in price according to size and finish from 12*s.* 6*d.* up to 26*s.*

Small-power engines. We meanwhile leave these

engines out of account, but have no hesitation in saying that the amateur who has successfully made any of the models mentioned here, may safely take in hand to make a small-power engine, suitable for driving his lathe, from a set of castings as supplied by the model-makers. If the cylinder is bored before the castings are sent out, the engine will be easily made. This had best be done, for the cylinder requires very careful workmanship. A suitable boiler, with all fittings, should be purchased ready-made. We do not advise any one to attempt small-power boiler-making at home, as the result can hardly be anything but disappointing, and probably disastrous. An engineer, or even a boiler-maker, can do very little work at home, consisting merely of riveting up, and attaching fittings and mountings. Rolls are required for bending the plates, a furnace for heating, and a templet-block for flanging the crowns, the cross-tube, and for welding up, and this can only be done successfully in an engineering workshop.

NOTES.

WHEN water boils (at a temperature of 212° Fahr.) a vapour is given off which is called steam. This possesses, like other gases, two properties, viz. expansibility and elasticity, along with another, condensability, all of which render it of value in imparting motion to a steam-engine.

Expansibility. This property can be shown by purchasing one or two small glass candle bombs, which contain water, and when one of them is stuck by its narrow end into the top of a tallow candle at the side of the wick and lighted, in a short time it explodes with a report into innumerable fragments. This result is caused by steam at atmospheric pressure occupying 1642 times the volume of its weight of water, or, in other words: water occupying the space of a cubic inch when converted into steam will occupy the space of a cubic foot; and suppose the bomb held the above amount of water, consequently, when this water becomes steam there is no room for it in the bomb, and the pressure increases till it shatters the bomb all to pieces.

Condensability. This is the property which steam possesses of returning to water when the heat is suddenly withdrawn, and a lowering of temperature takes place. Advantage was taken of this fact in the early forms of the Newcomen atmospheric beam engine, where the weight of the beam raised the piston to the top of the cylinder, and by suitable arrangements the steam space below the piston was then filled with steam, which was suddenly condensed by a jet of cold water. This setting up a vacuum, caused the piston to descend, as it was forced down by atmospheric pressure, and so completed the stroke.

Elasticity. This is the property steam possesses of having the power of resistance to external pressure, and the steam which is formed at 212° Fahr. possesses just that degree of elastic force which is required to balance and resist the pressure of the atmosphere. At 212° Fahr. its elastic force just equals one atmosphere, or 14·7 lbs. per square inch. This property, and the previous one mentioned above, can be easily shown by Wollaston's apparatus, which consists of a bulb and glass tube provided with a piston and hollow piston-rod working freely through a cork. On filling the bulb with water and heating it, so long as the cock communicating with the piston-rod remains open, there is an escape of steam; on closing the cock, steam forces the piston upwards, subsequent cooling causes condensation of the steam, and the piston descends by atmospheric pressure to its former position.

Saturated steam is steam in contact with the water from which it is generated. This is the condition in which it is usually supplied to engines.

Super-heated steam is steam absolutely dry, and with no vapoury particles held in suspension; its temperature is also higher than that due to the corresponding pressure of saturated steam.

Horse-power of an engine.

Energy is the capability of doing work, and this is measured by the number of units of work done. Power is the rate of working, or the work done in unit time; so, according to engineers, the unit of work is the foot pound, and the unit of power is called a horse-power (H.P.), which is the power of doing 33,000 foot pounds of work per minute.

The horse-power is calculated as follows:—

Let A equal the area of the piston in square inches, P the pressure in lbs. per square inch of steam, L the length of stroke in feet per minute, and N the number of revolutions per minute, then

$$\frac{A \times P \times L \times N}{33,000} = \frac{\cdot 196350 \times 30 \times \cdot 083 \times 400}{33,000} = \cdot 059 \text{ H.P.}$$

An engine with a cylinder of $\frac{1}{2}''$ bore and $1''$ stroke, working at 30 lbs. pressure, and making 200 revolutions per minute (by our calculation), will develop about $\frac{1}{16}$th H.P.; if the mean pressure of steam was taken, and allowance made for friction, probably it would not amount to $\frac{1}{32}$nd H.P.

In a model locomotive, steam of a greater pressure than about two atmospheres, or 30 lbs. per square inch, cannot be well maintained, hence it is advisable never to make the locomotive of too heavy material, but have it sufficiently strong only to be consistent with durability, and then it will give the best results when running upon rails.

INDEX.

LOCOMOTIVE BOILER.—SHEET NO. 1.

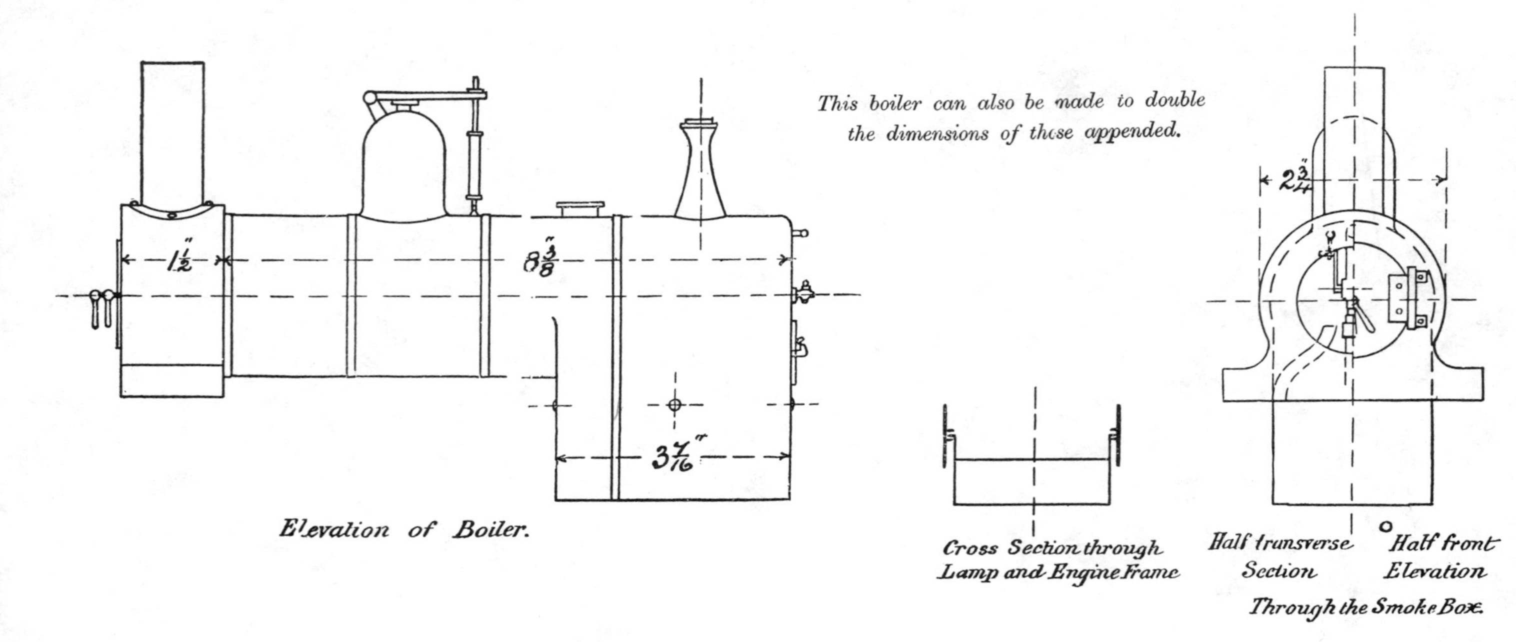

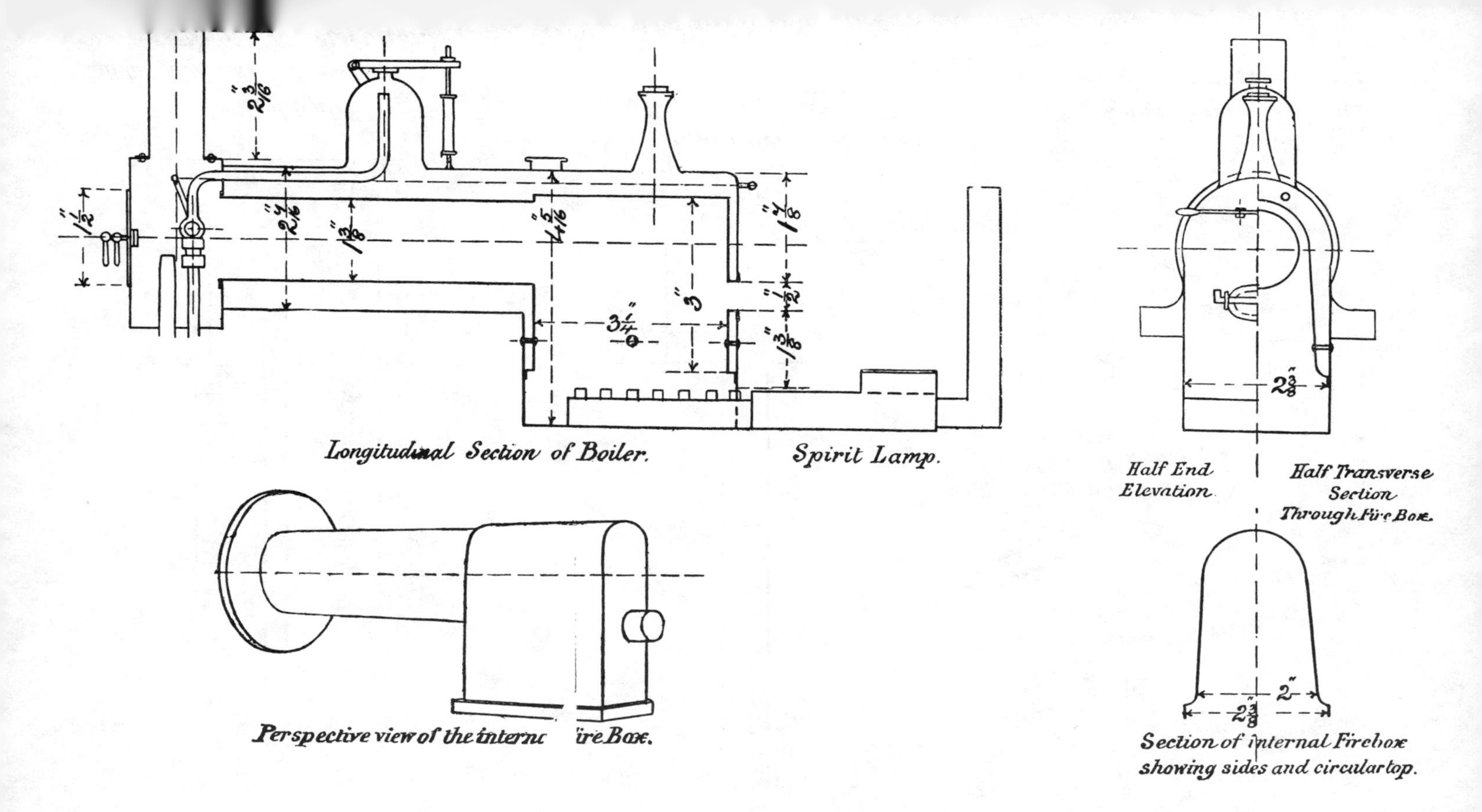
2 3/16"
1/2"
2 1/16"
1 3/8"
4 5/16"
3"
3 1/4"
1 1/8"
1/2"
1 3/8"
Longitudinal Section of Boiler.
Spirit Lamp.
2 3/8"
Half End Elevation.
Half Transverse Section Through Fire Box.
Perspective view of the interna ire Box.
2"
2 3/8"
Section of internal Firebox showing sides and circular top.

VERTICAL BOILER.—SHEET NO. 2.

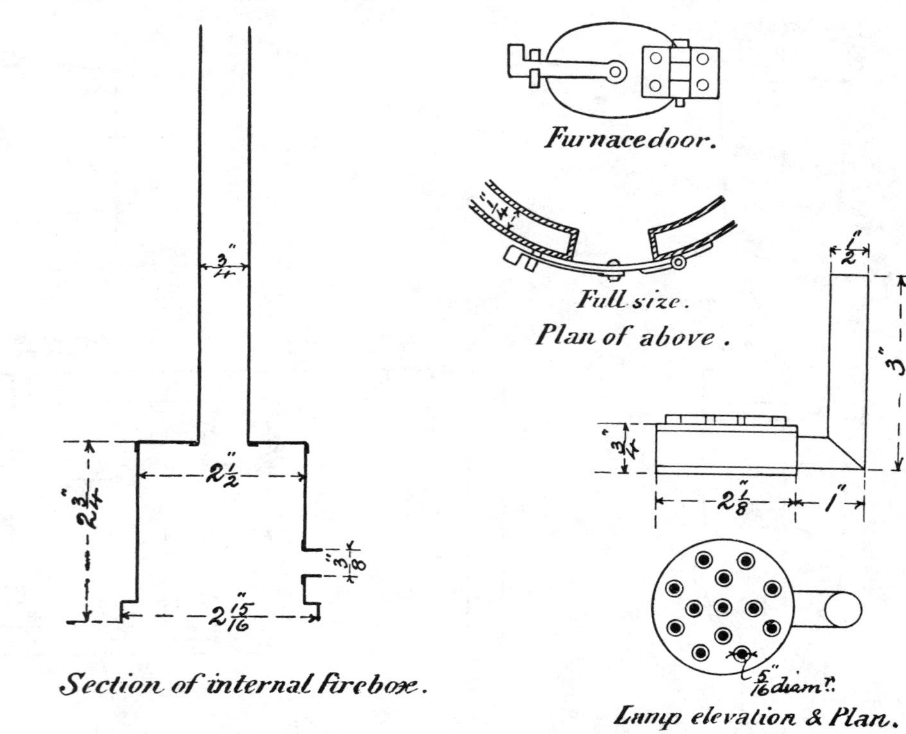

BEAM ENGINE.—SHEET NO. 3.

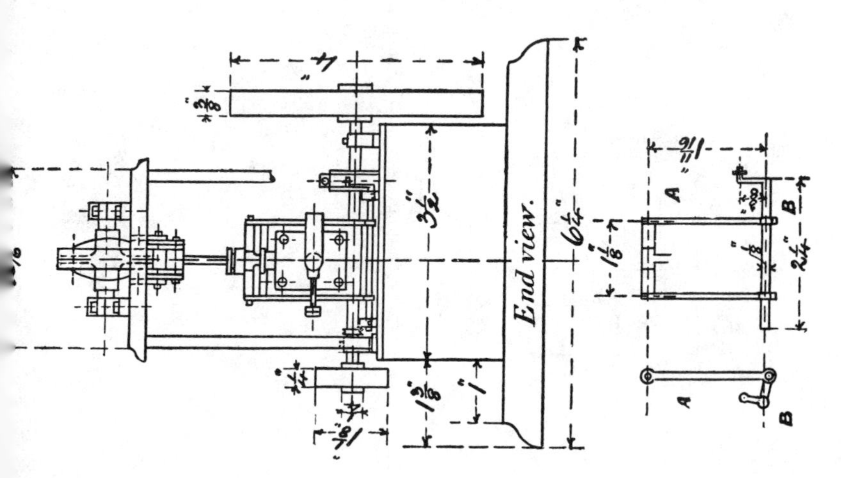

Levers to move slide valve spindle.

A, Vertical rod. B, Bell-crank.

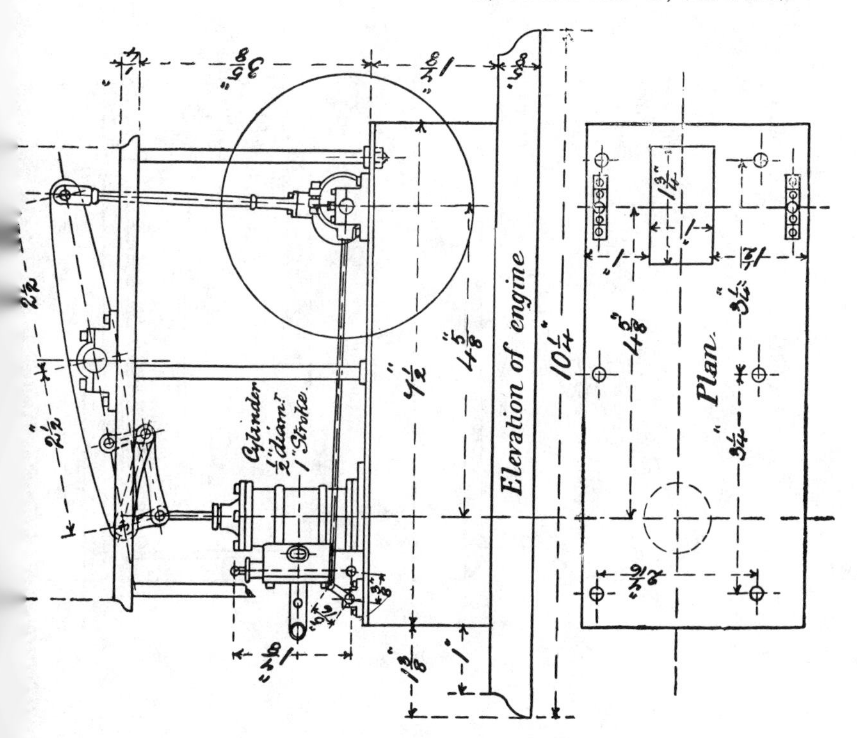

HORIZONTAL ENGINE.—SHEET NO. 4.

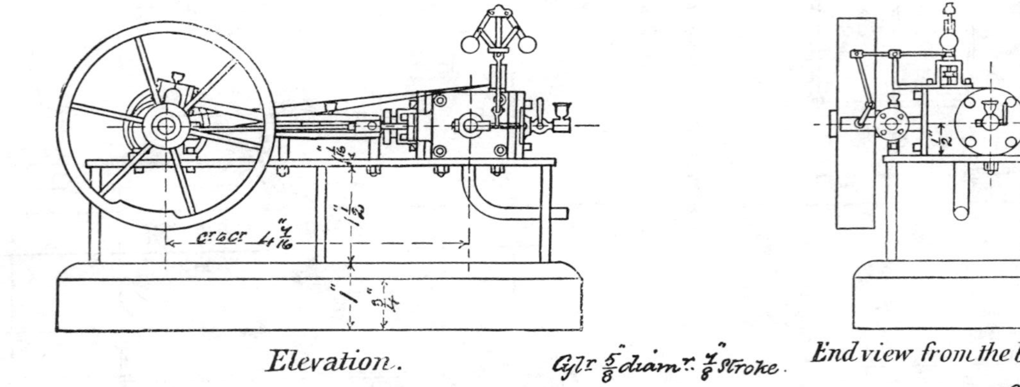

Elevation.

Cylr. 5/8" diamr. 7/8" Stroke.

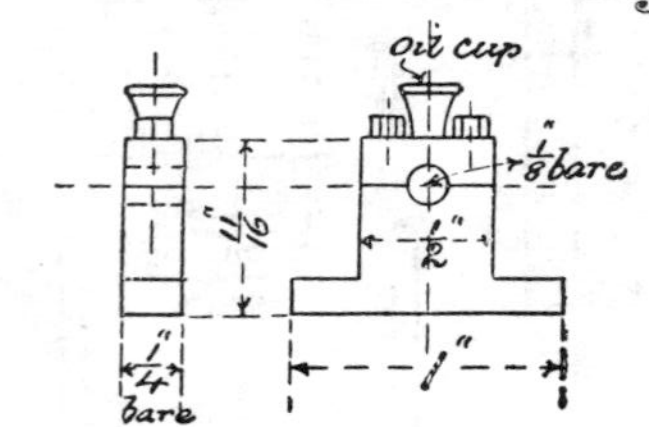

End view from the back of Engine.

Bearings.

Full size.

HORIZONTAL ENGINE.—SHEET NO. 5.

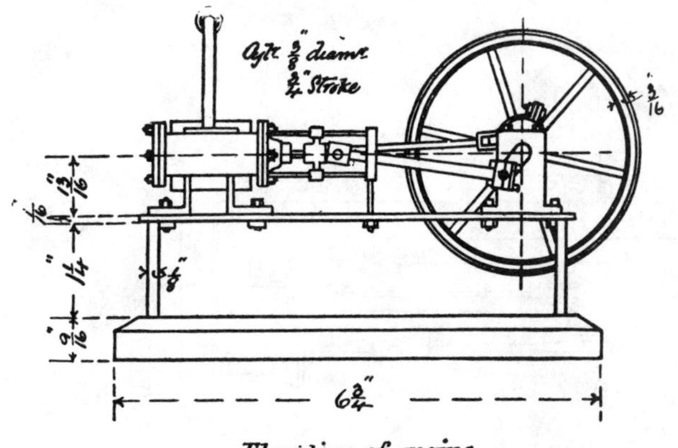

Elevation of engine.

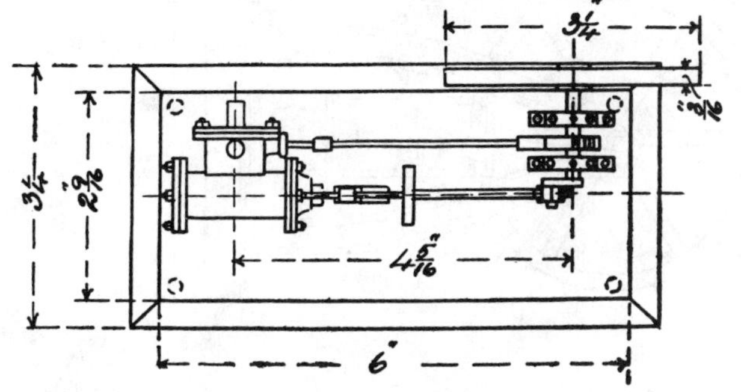

Plan of engine.

VERTICAL ENGINE.—SHEET NO. 6.

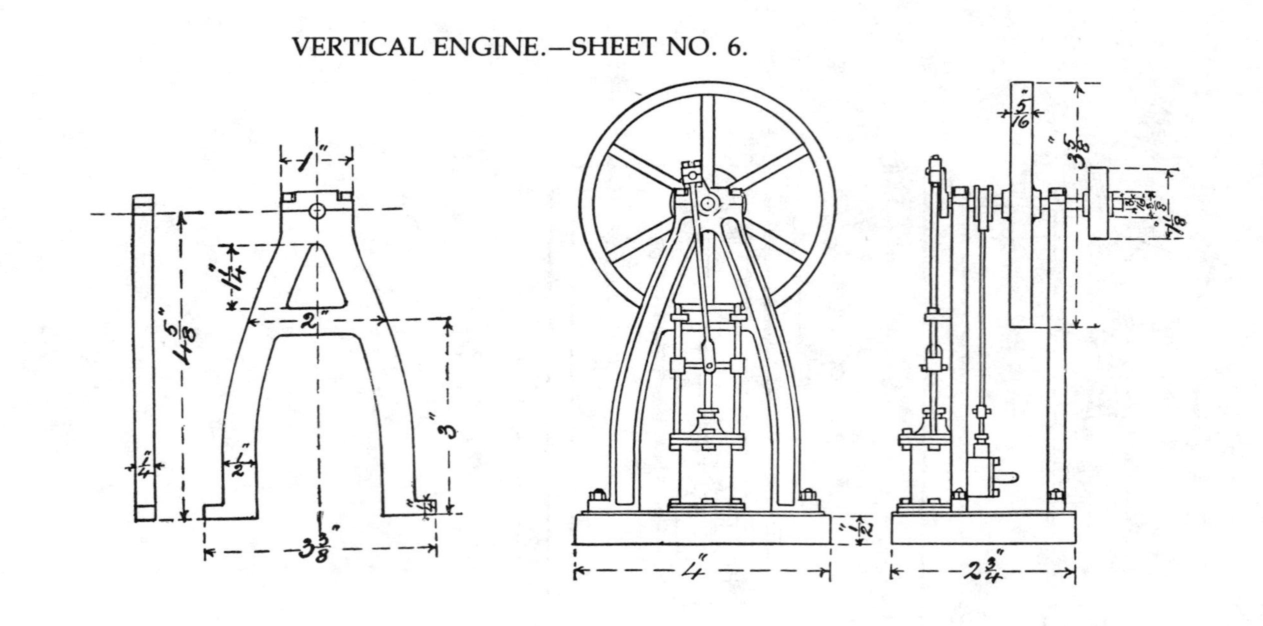

INVERTED VERTICAL ENGINE.—SHEET NO. 7.

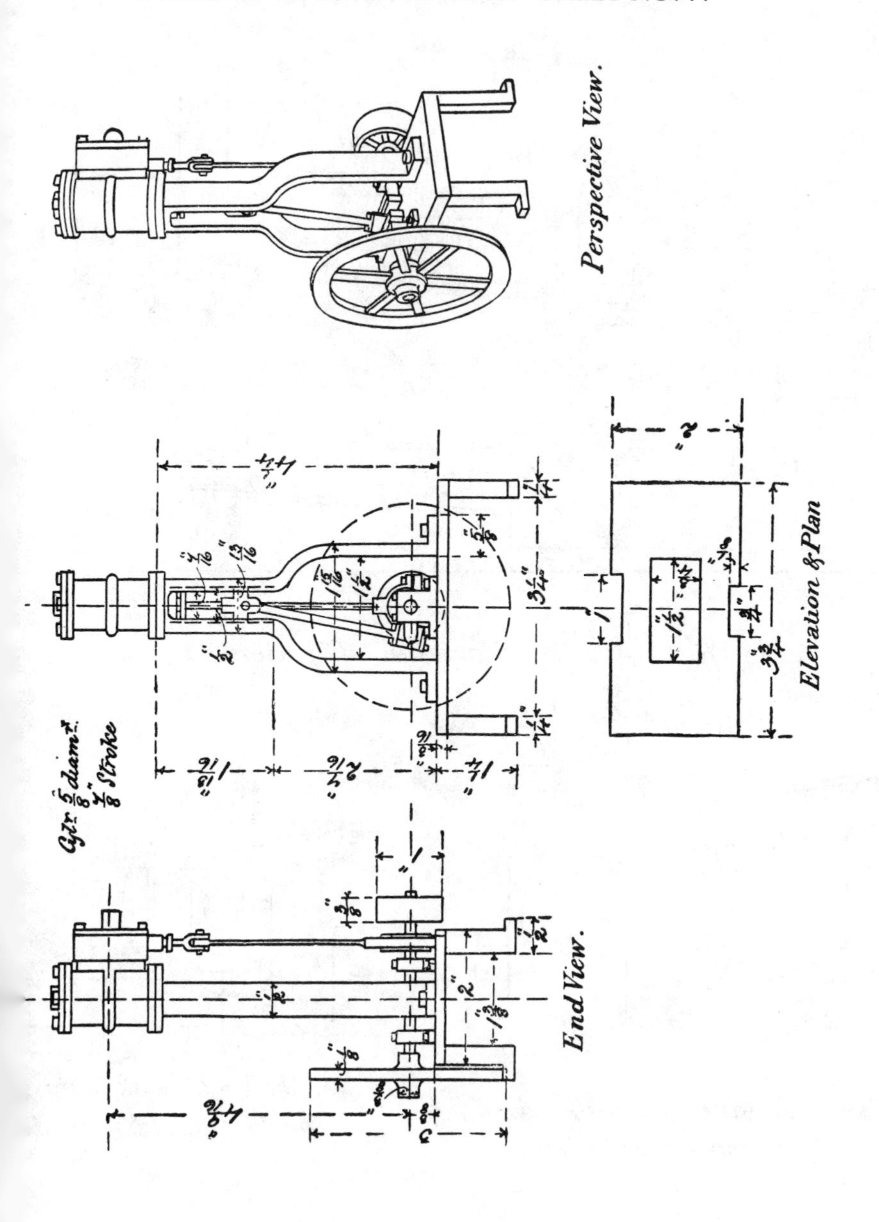

TRACTION ENGINE.—SHEET NO. 8, A.

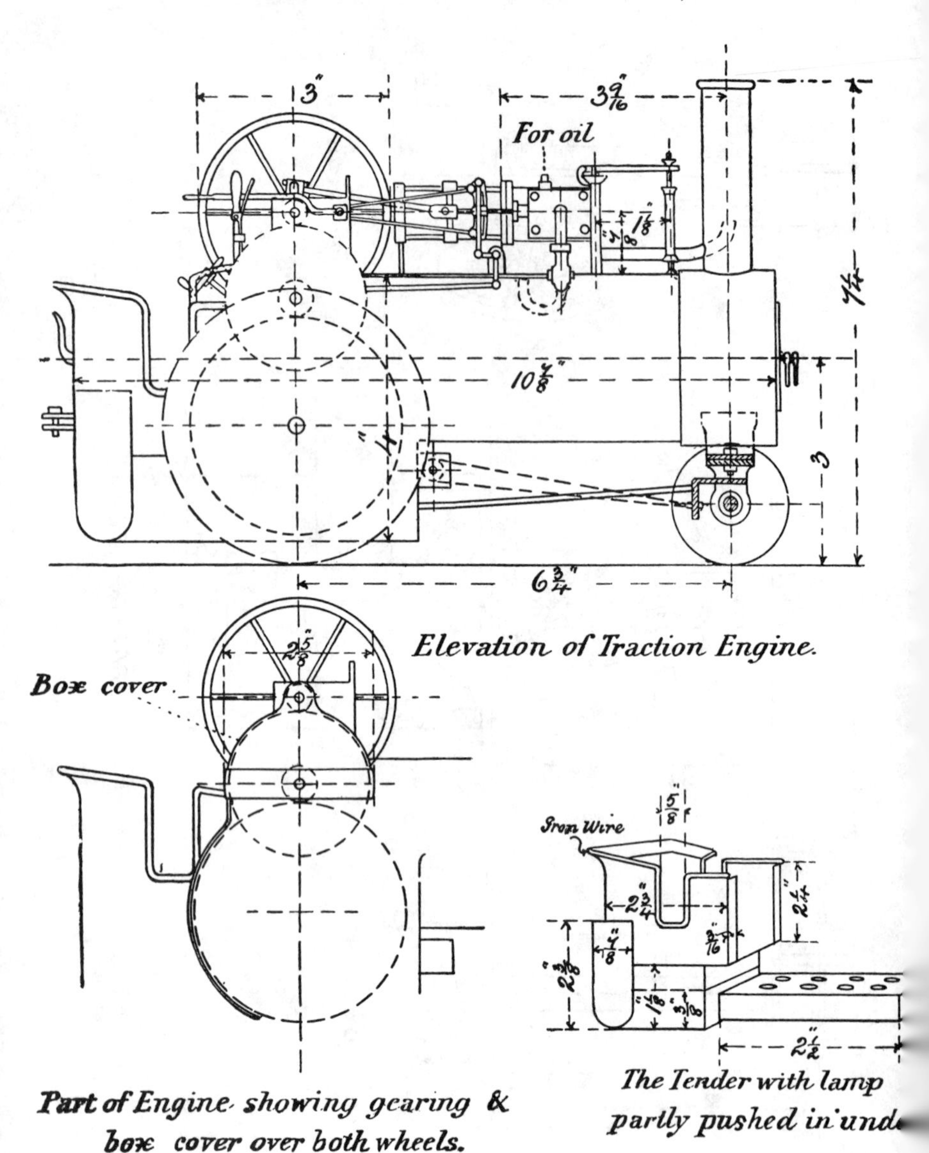

Elevation of Traction Engine.

Part of Engine showing gearing & box cover over both wheels.

The Tender with lamp partly pushed in unde

TRACTION ENGINE.—SHEET NO. 8, A.

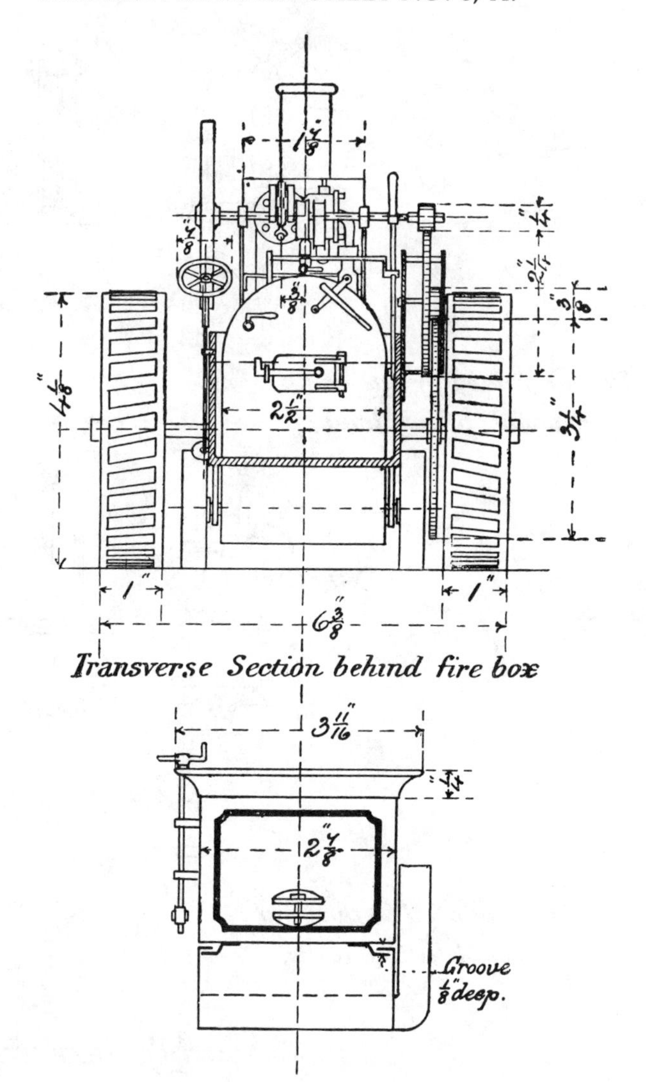

Transverse Section behind fire box

Back of Tender with Spirit Lamp

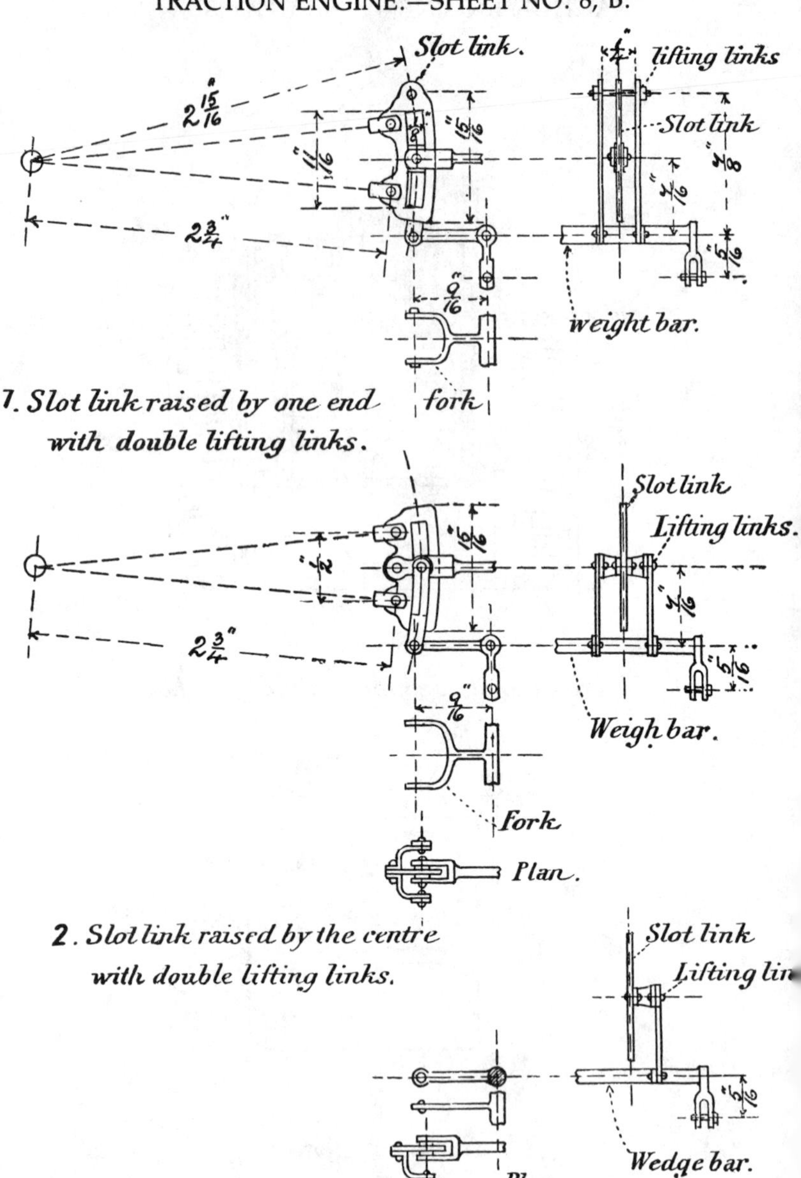

1. Slot link raised by one end with double lifting links.

2. Slot link raised by the centre with double lifting links.

3. Slot link raised by the centre with a single lifting link only.

TRACTION ENGINE.—SHEET NO. 8, B.

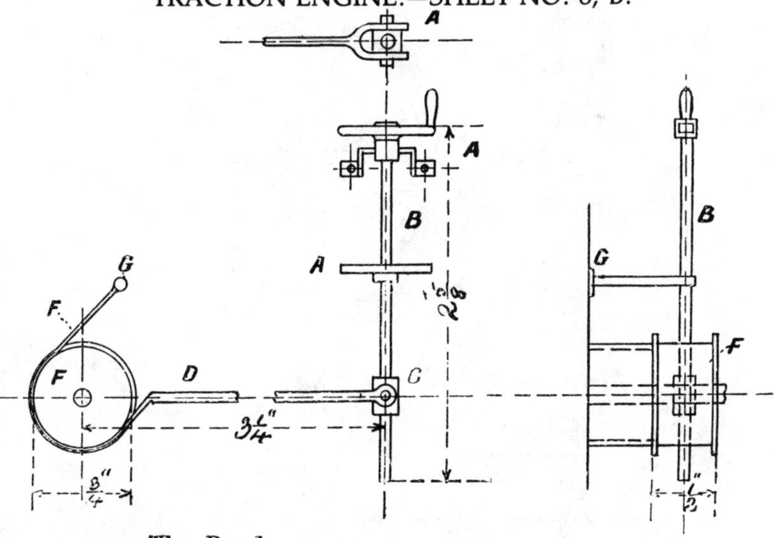

The Brake.

A. Brackets. B. Brake rod. C. Pivot nut.
D. Lever. E. Strap going round pulley F.
& attached to fulcrum G.

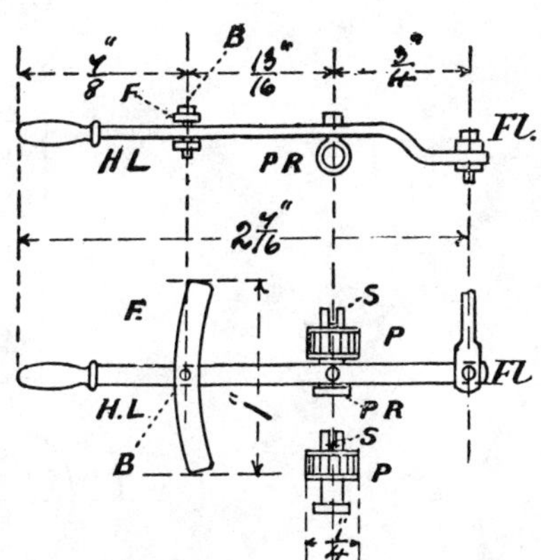

Clutch for throwing upper pinion into gear.

P. Pinion with boss. F. Fork. Fl. Fulcrum.

P. R. Pivoted ring. H. L. Hand-lever.

S. Slot in boss of pinion.

B. Pin to hold lever fixed in fork.

TRACTION ENGINE.—SHEET NO. 8, C.

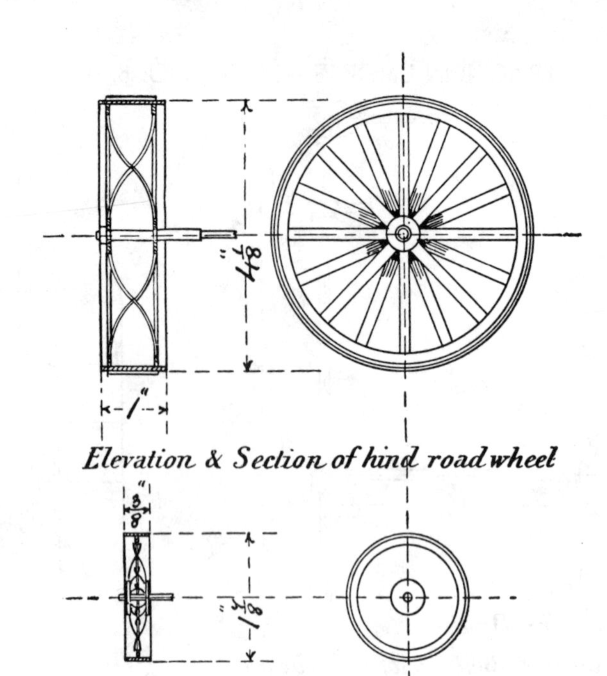

Elevation & Section of hind road wheel

Elevation & Section of leading wheel.

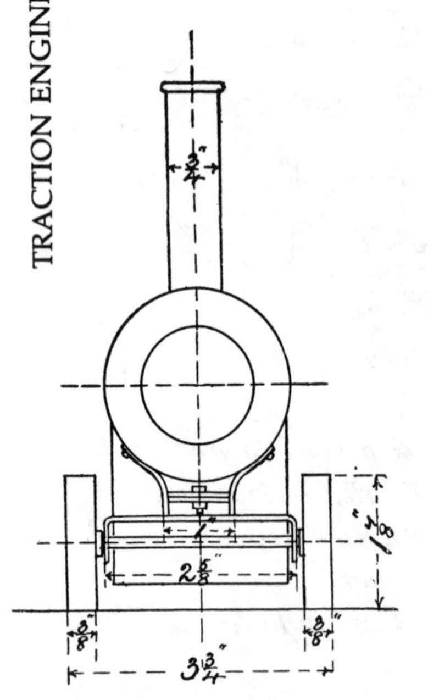

Front Elevation

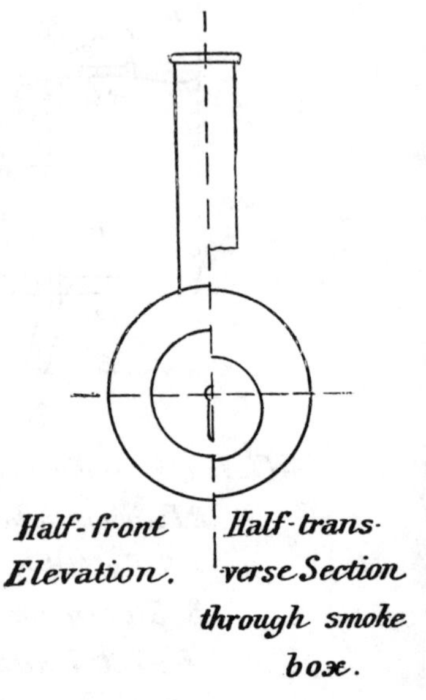

Half-front Elevation.

Half-transverse Section through smoke box.

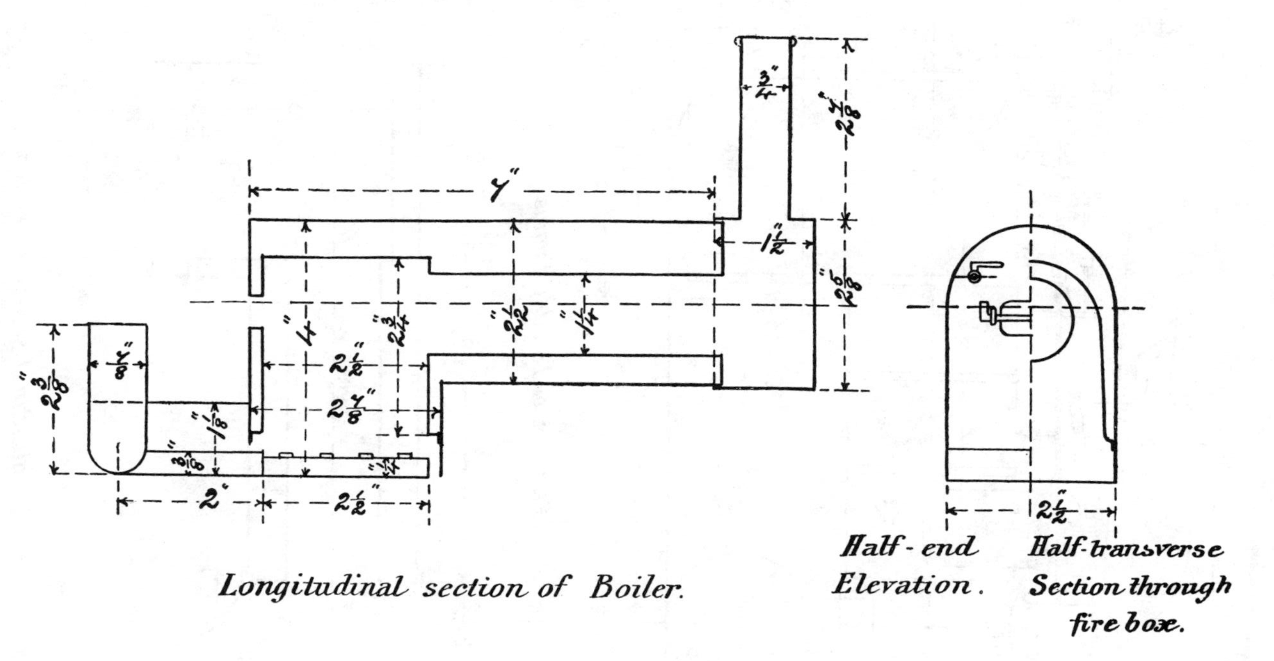

Longitudinal section of Boiler.

Half-end Elevation.

Half-transverse Section through fire box.

PORTABLE ENGINE.—SHEET NO. 9.

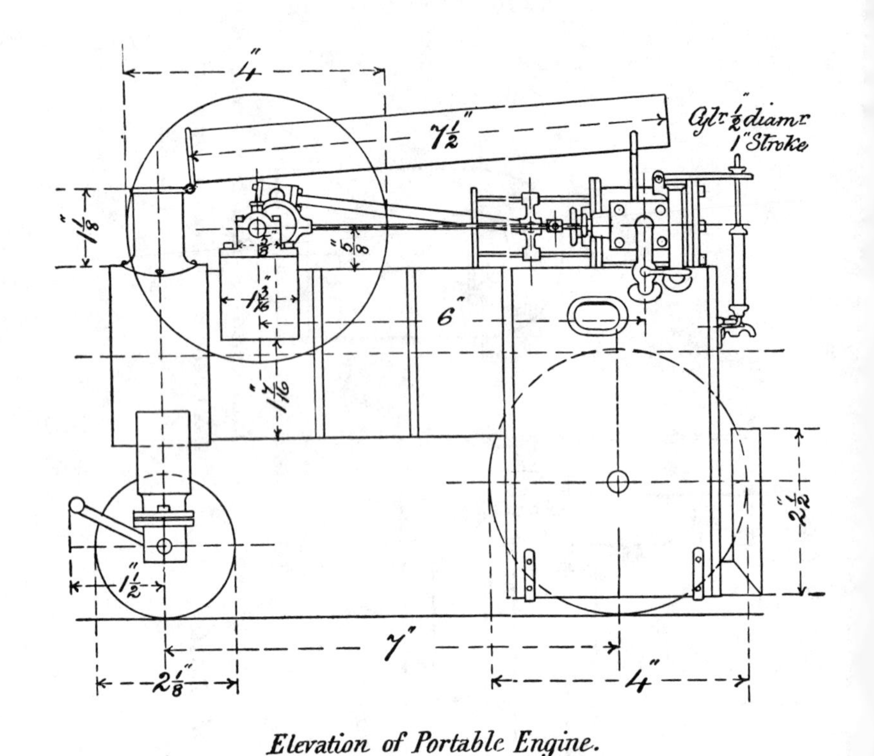

Elevation of Portable Engine.

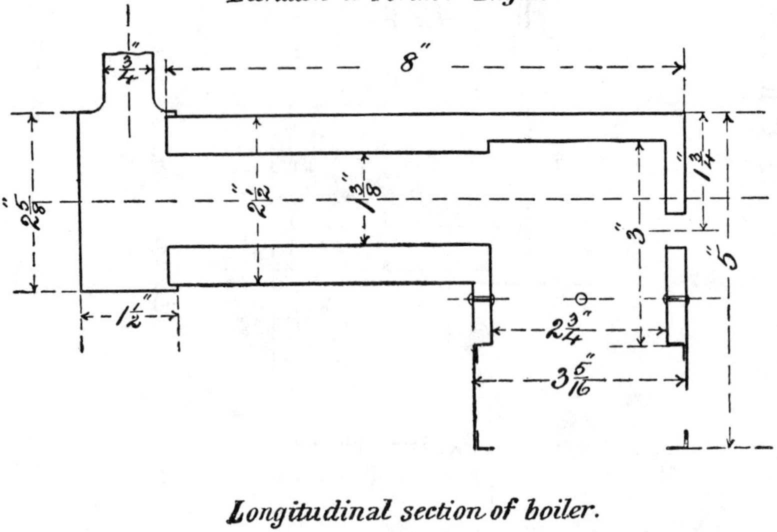

Longitudinal section of boiler.

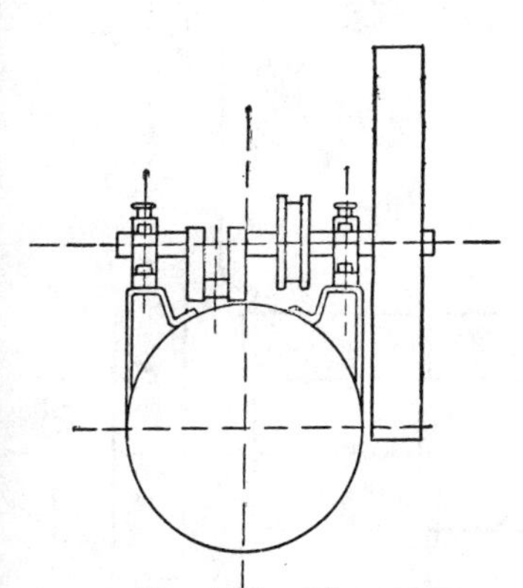

Front Elevation showing the bearings supported on brackets attached to the boiler.

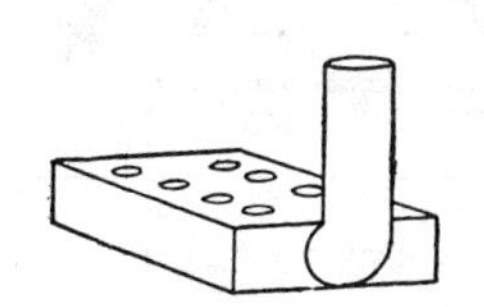

Perspective view of lamp.

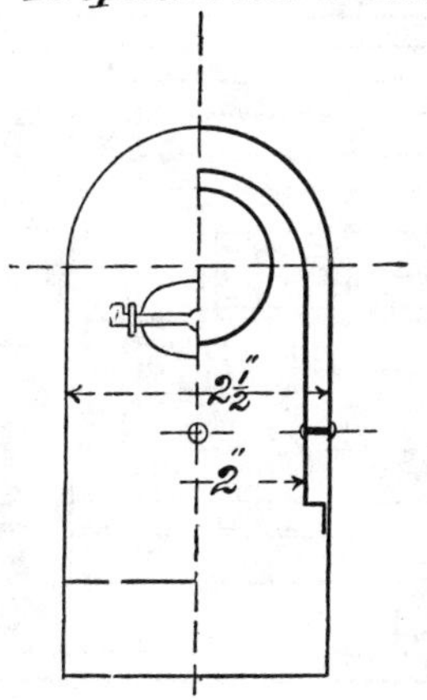

Half-end elevation.   *Half-transverse Section through firebox.*

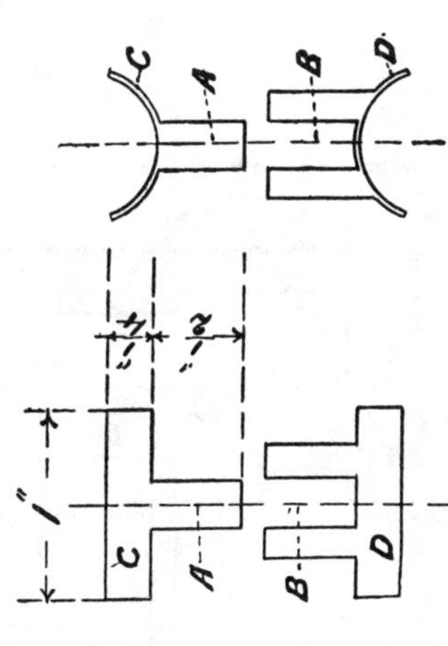

Fig. 1. **Snug A must fit within slot B when put together C & D are flat.**

Fig. 2. **C & D are bent into a semicircle. A & B are turned at right angles to C & D**

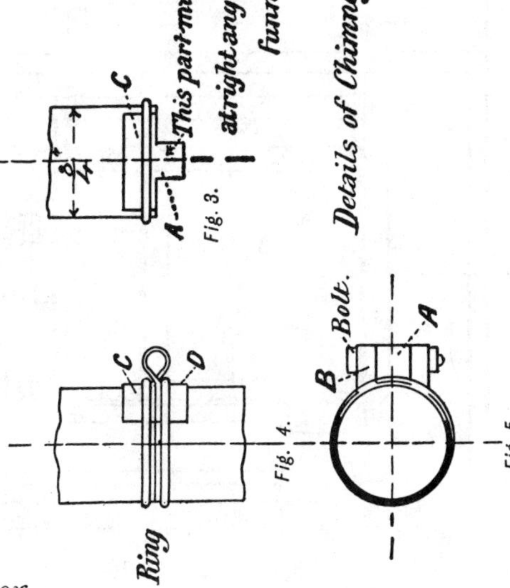

SEMI-PORTABLE ENGINE.—SHEET NO. 10.

This engine can also be made to double the dimensions of those appended.

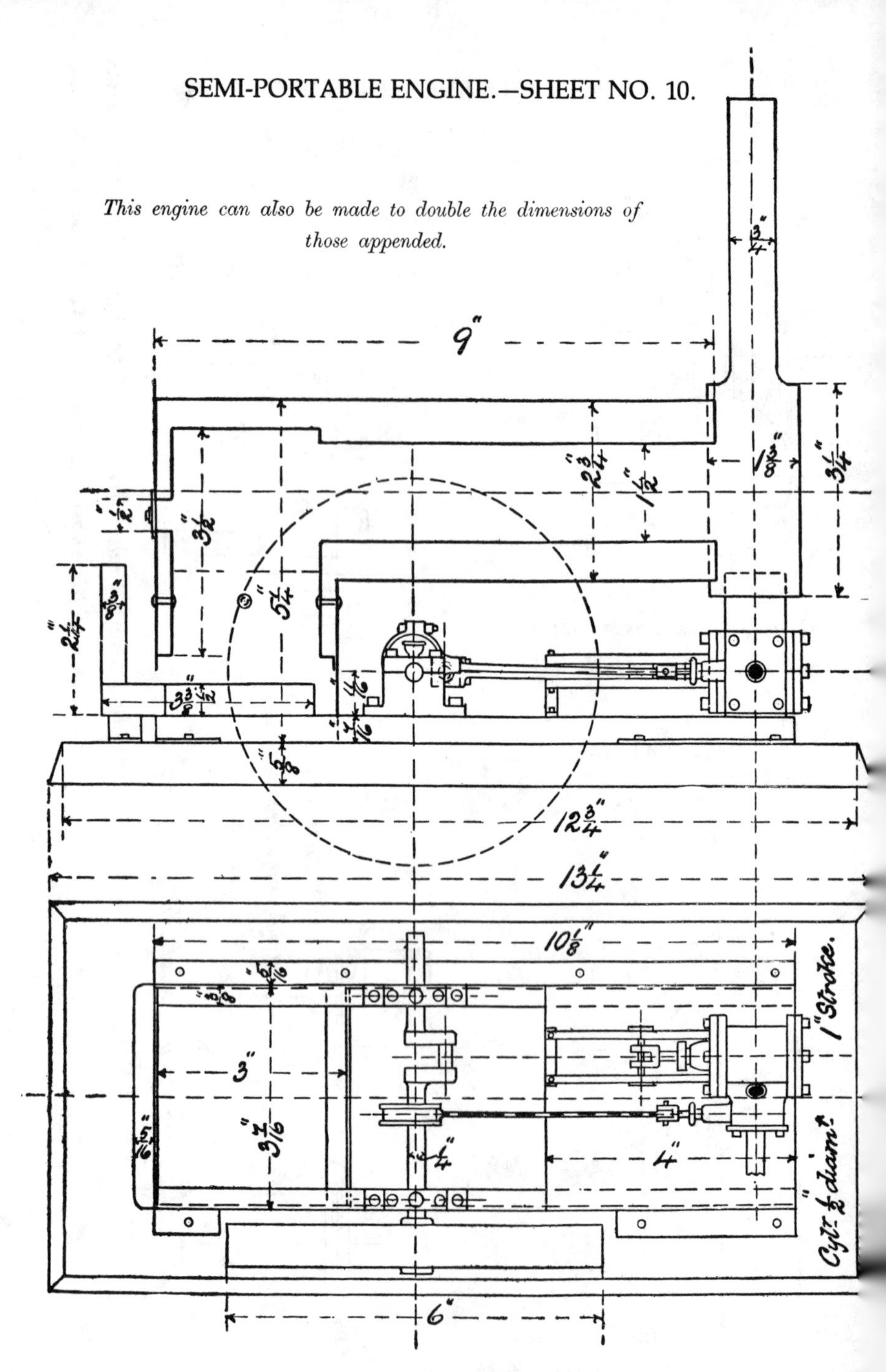

Longitudinal elevation & plan of Engine.

SEMI-PORTABLE ENGINE.—SHEET NO. 10.

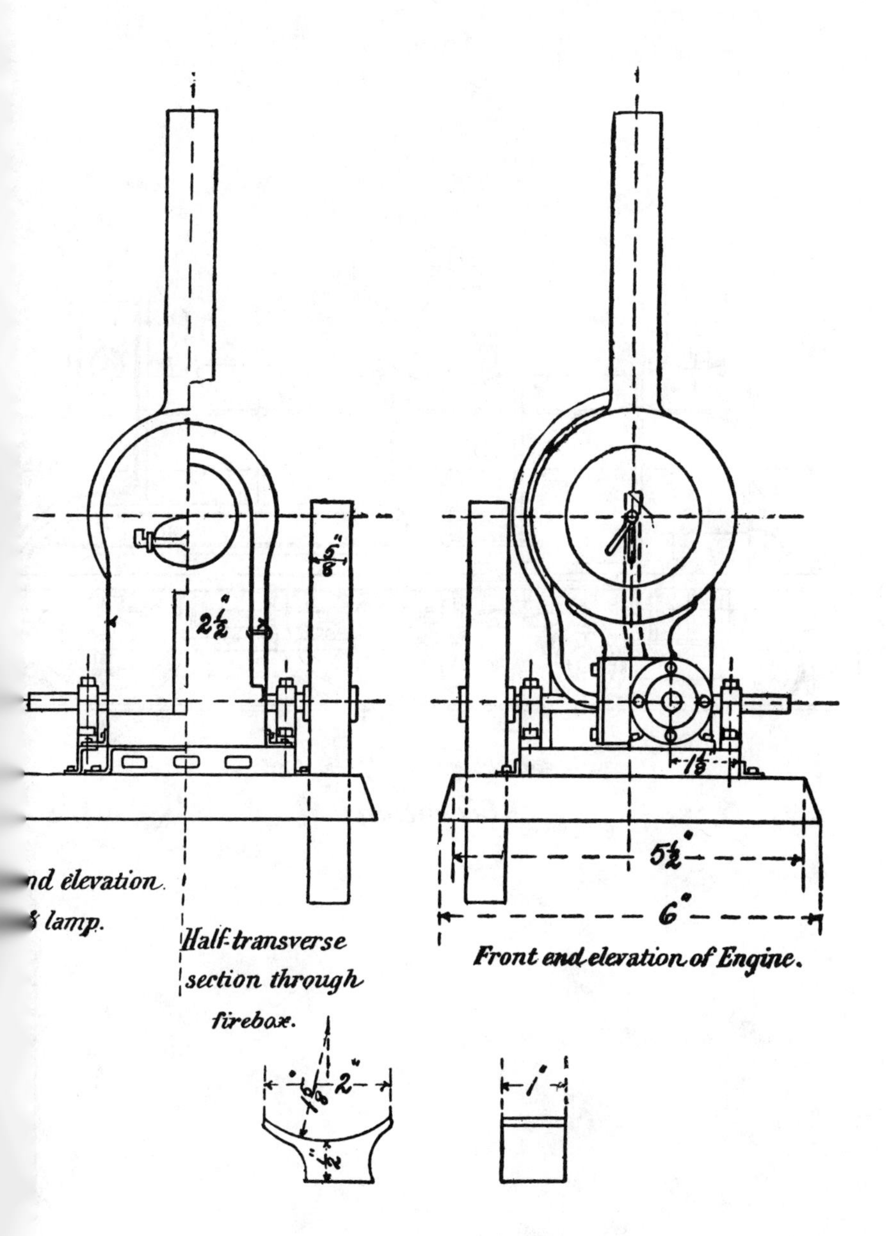

Elevation & plan of Saddle plate.

Total weight 7 lbs: 5 ozs.

Elevation.

LOCOMOTIVE AND TENDER.—SHEET NO. 11, A.

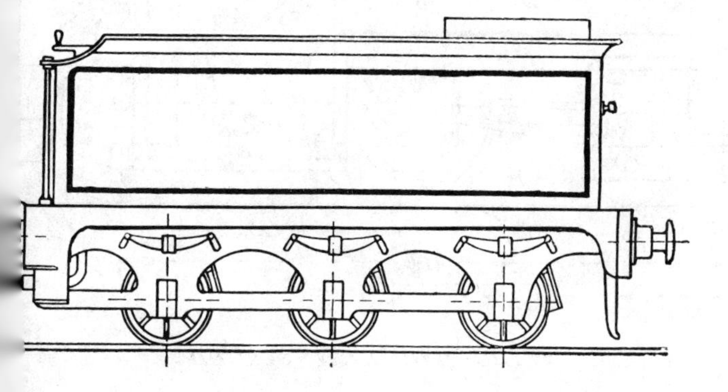

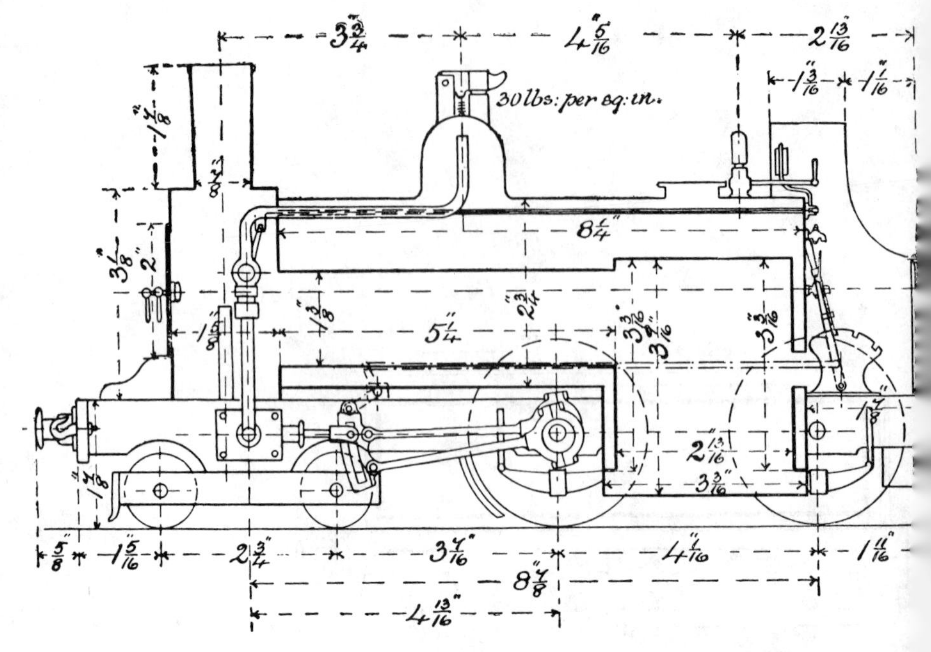

Longitudinal section.

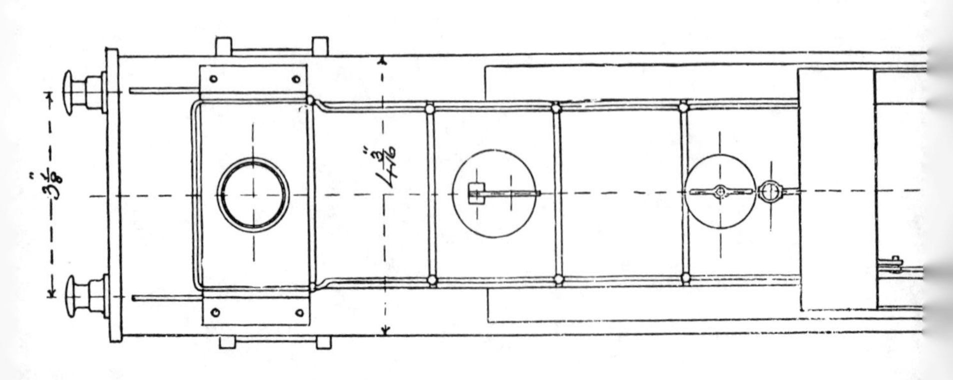

Plan of Locomotive.

LOCOMOTIVE AND TENDER.—SHEET NO. 11, A.

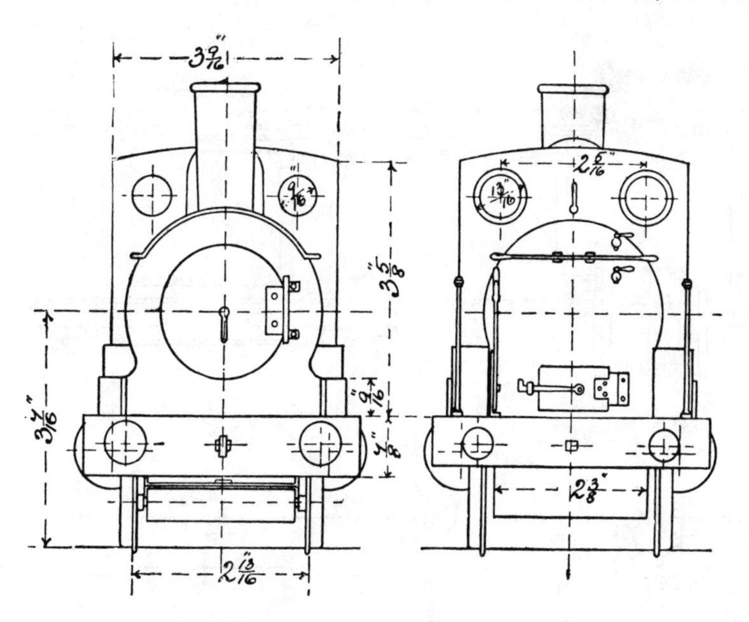

Front View. *Back View.*

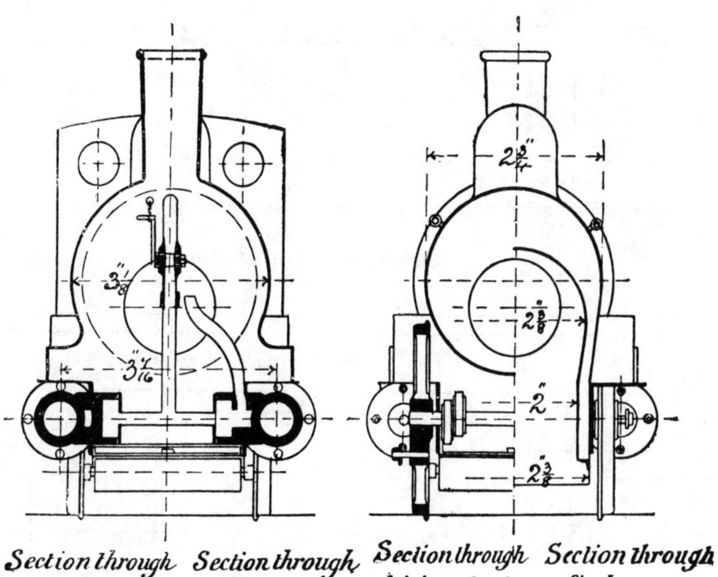

Section through steam pipe. *Section through exhaust pipe.* *Section through driving wheel.* *Section through firebox.*

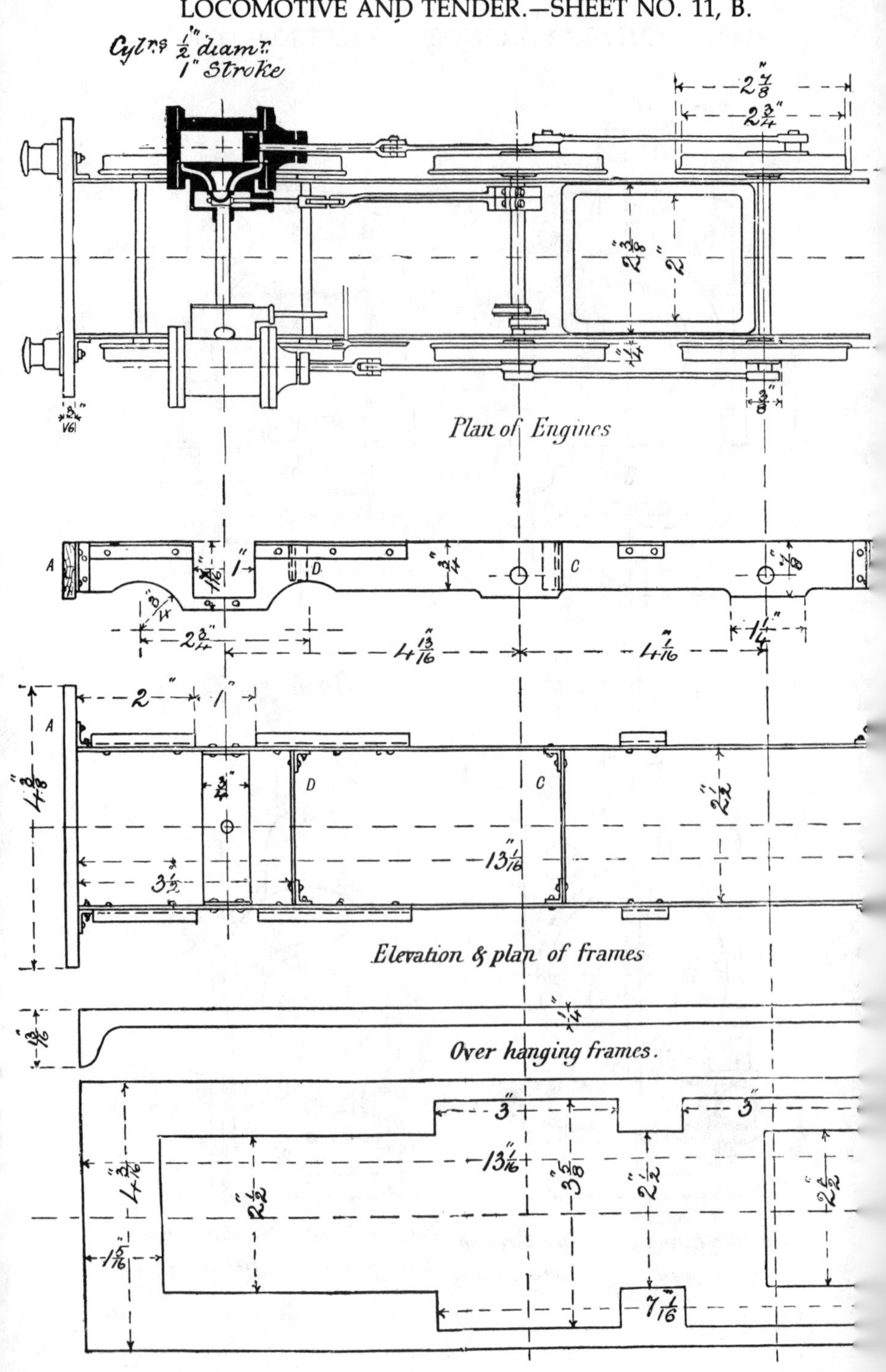
Cylrs ½" diamr. 1" Stroke
Plan of Engines
Elevation & plan of frames
Over hanging frames.

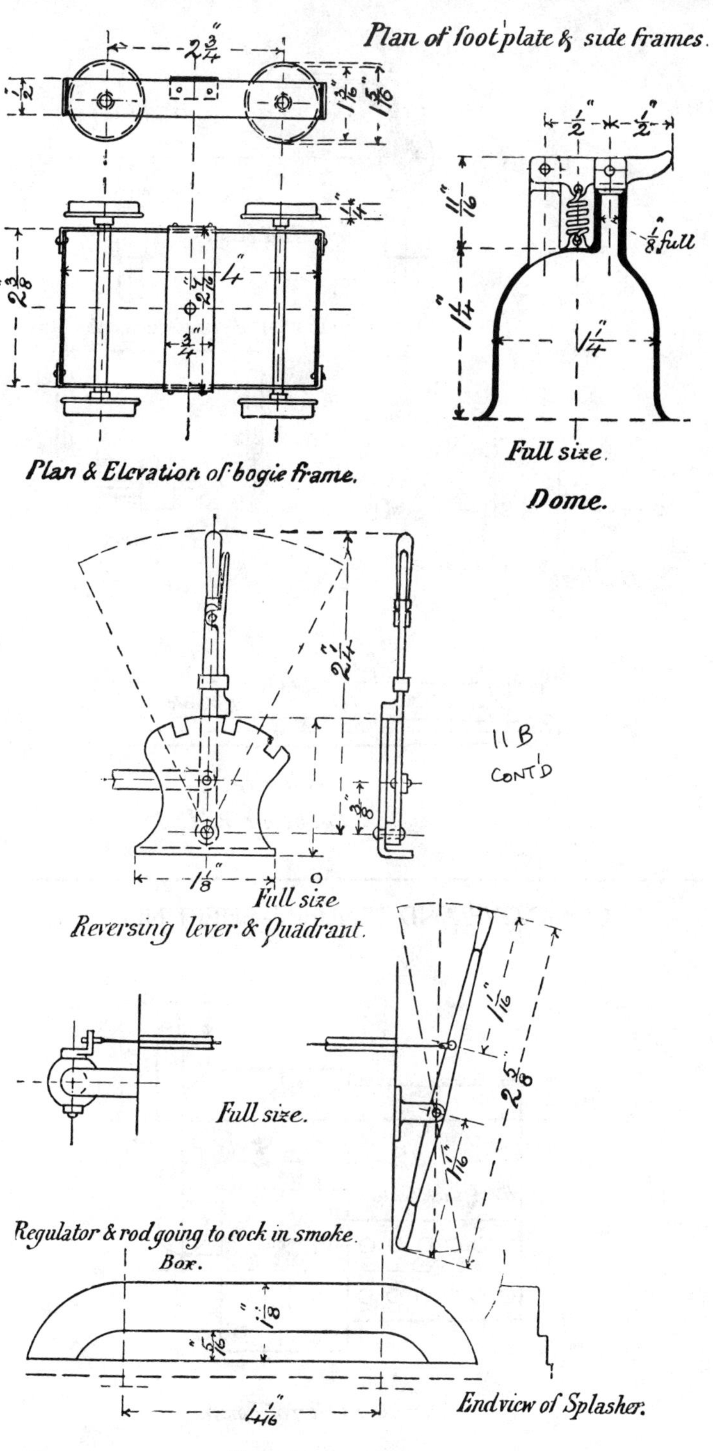
Plan of foot plate & side frames.
Plan & Elevation of bogie frame.
$\frac{1}{8}$ full
Full size.
Dome.
11 B
CONT'D
Full size
Reversing lever & Quadrant.
Full size.
Regulator & rod going to cock in smoke Box.
Endview of Splasher.
Splasher.

LOCOMOTIVE AND TENDER.—SHEET NO. 11, B.

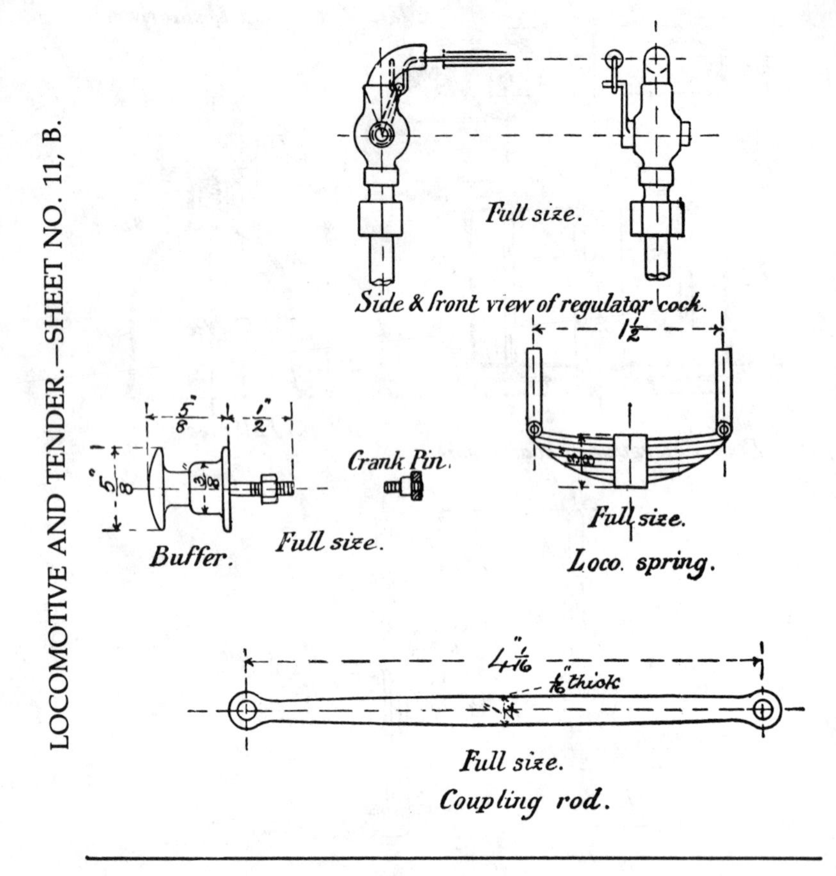

LOCOMOTIVE AND TENDER.—SHEET NO. 11, C.

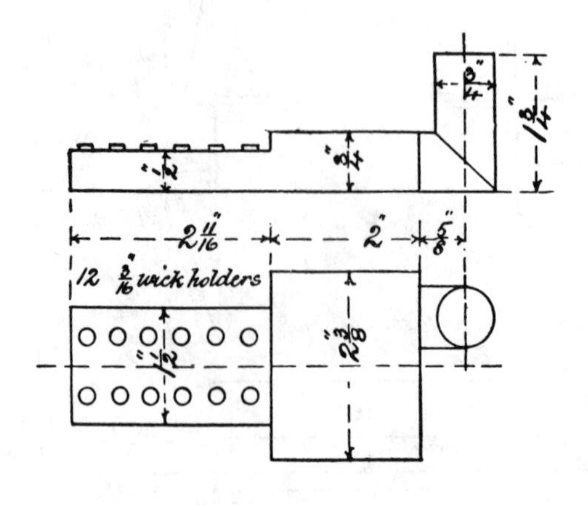

Elevation & Plan of Lamp.

LOCOMOTIVE AND TENDER.—SHEET NO. 11, C.

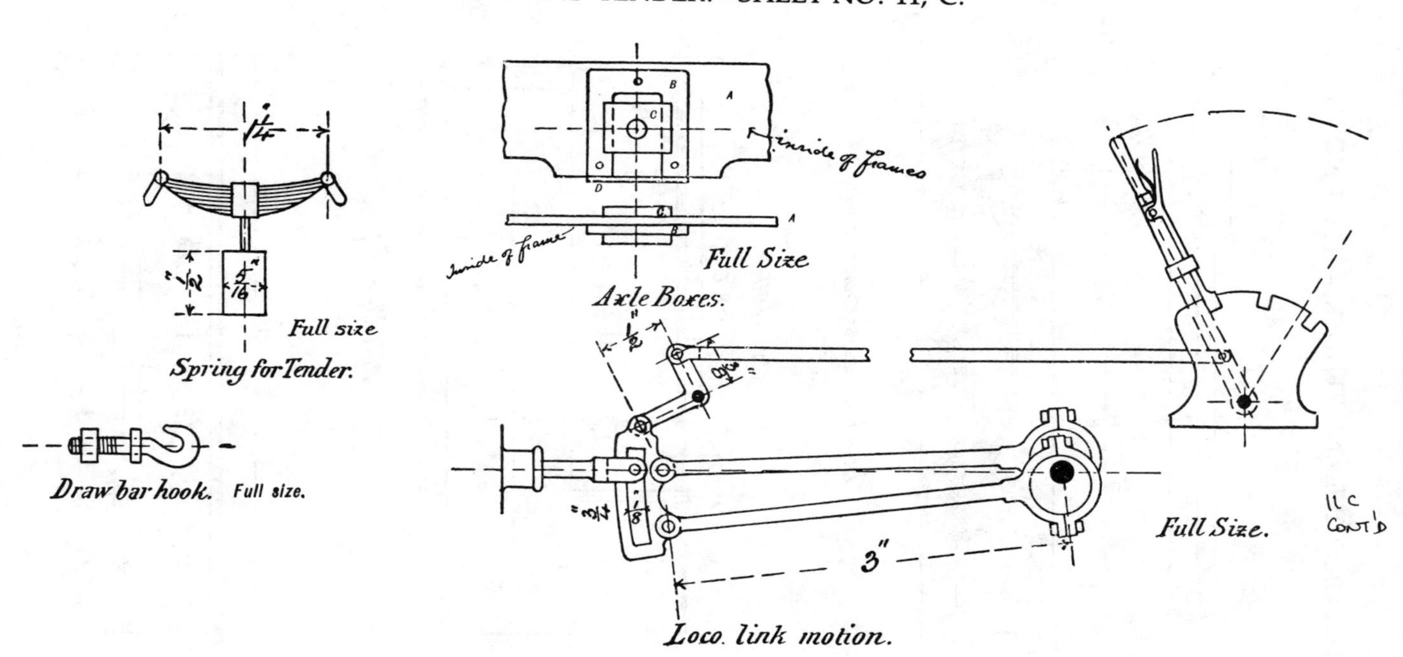

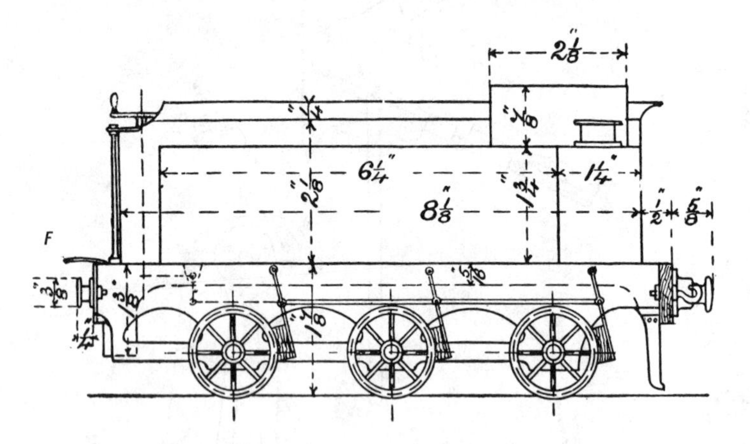

Longitudinal section of tender.

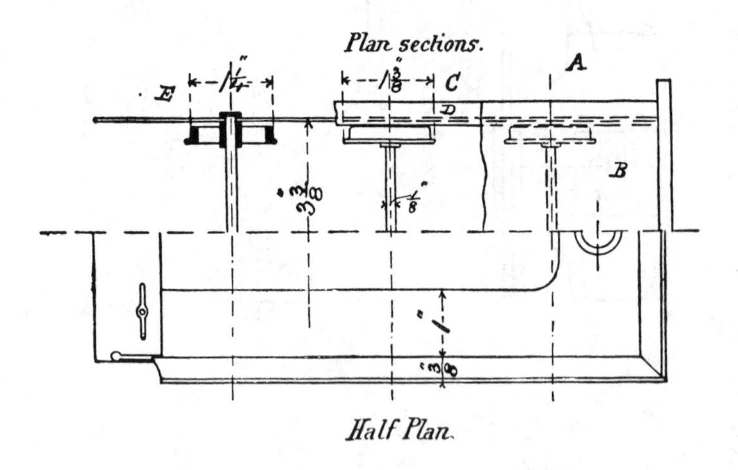

Half Plan.

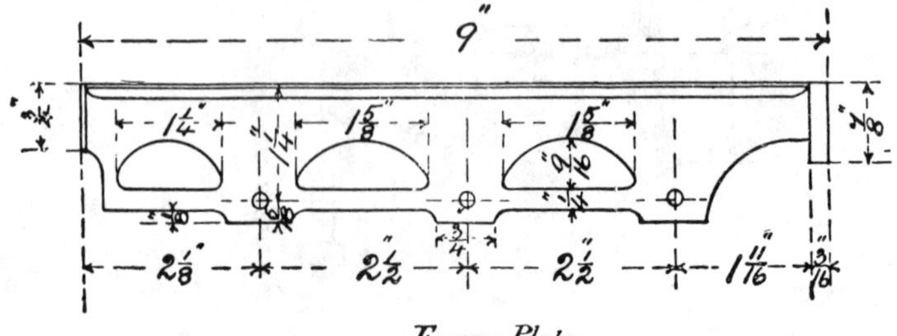

Frame Plate.

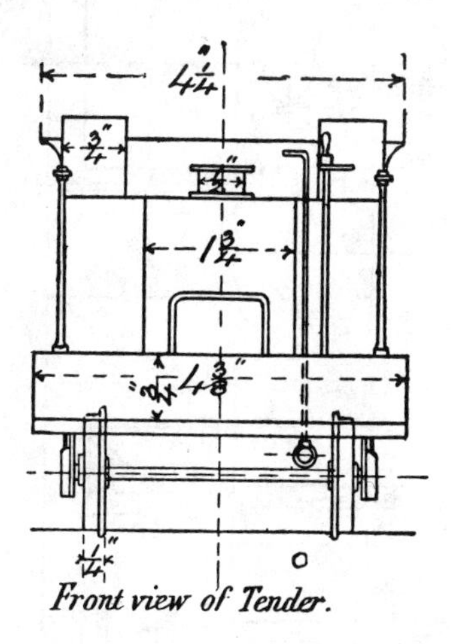

Front view of Tender.

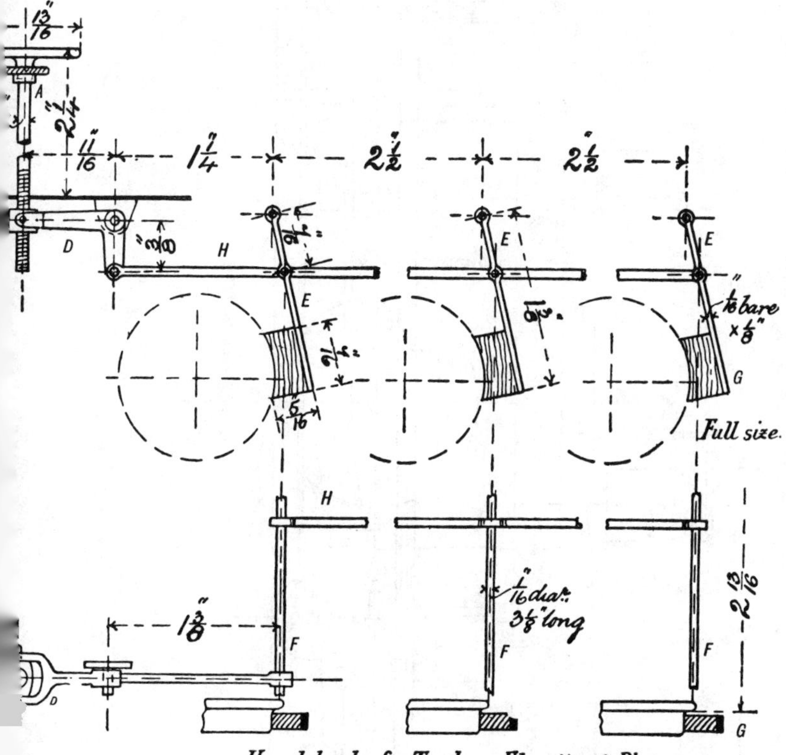

Hand brake for Tender. Elevation & Plan.

BOGIE-TANK LOCOMOTIVE.—SHEET NO. 12.

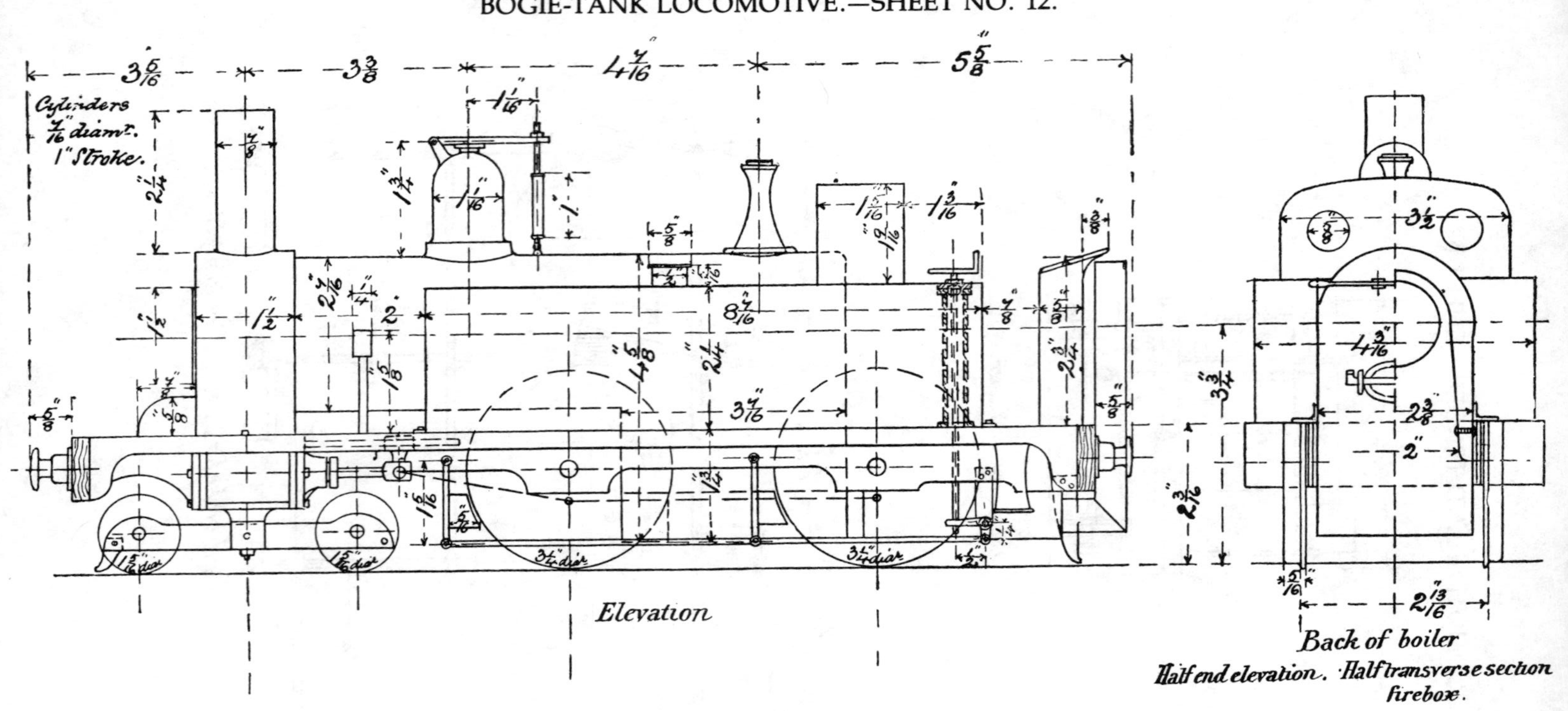

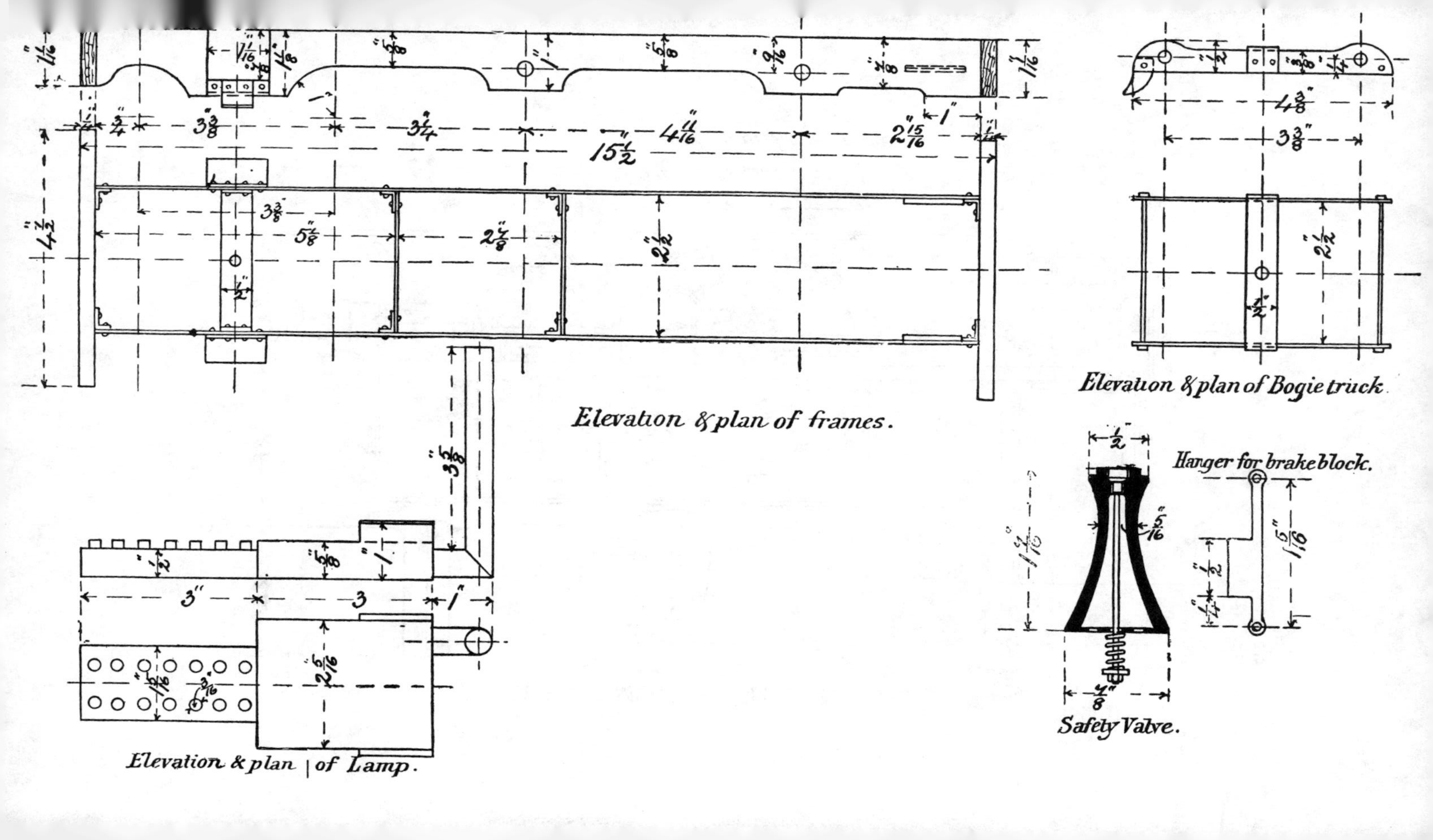
Elevation & plan of frames.
Elevation & plan of Bogie truck.
Hanger for brake block.
Safety Valve.
Elevation & plan | of Lamp.

SINGLE LOCOMOTIVE.—SHEET NO. 13.

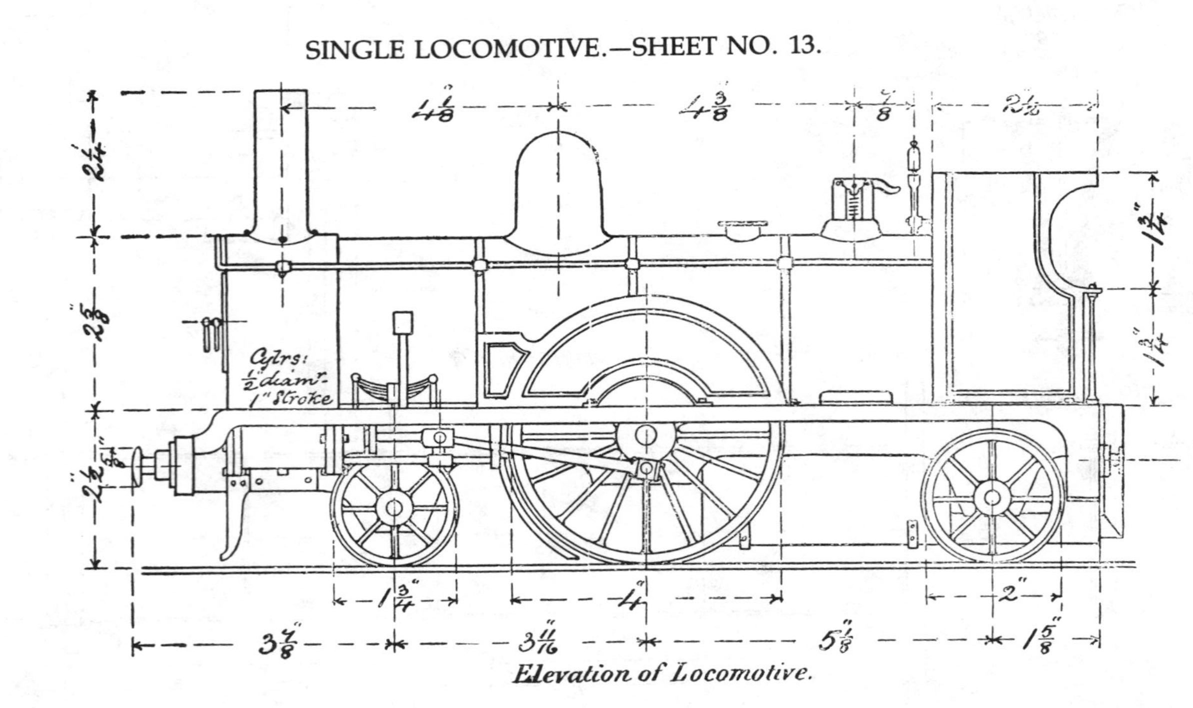

Elevation of Locomotive.

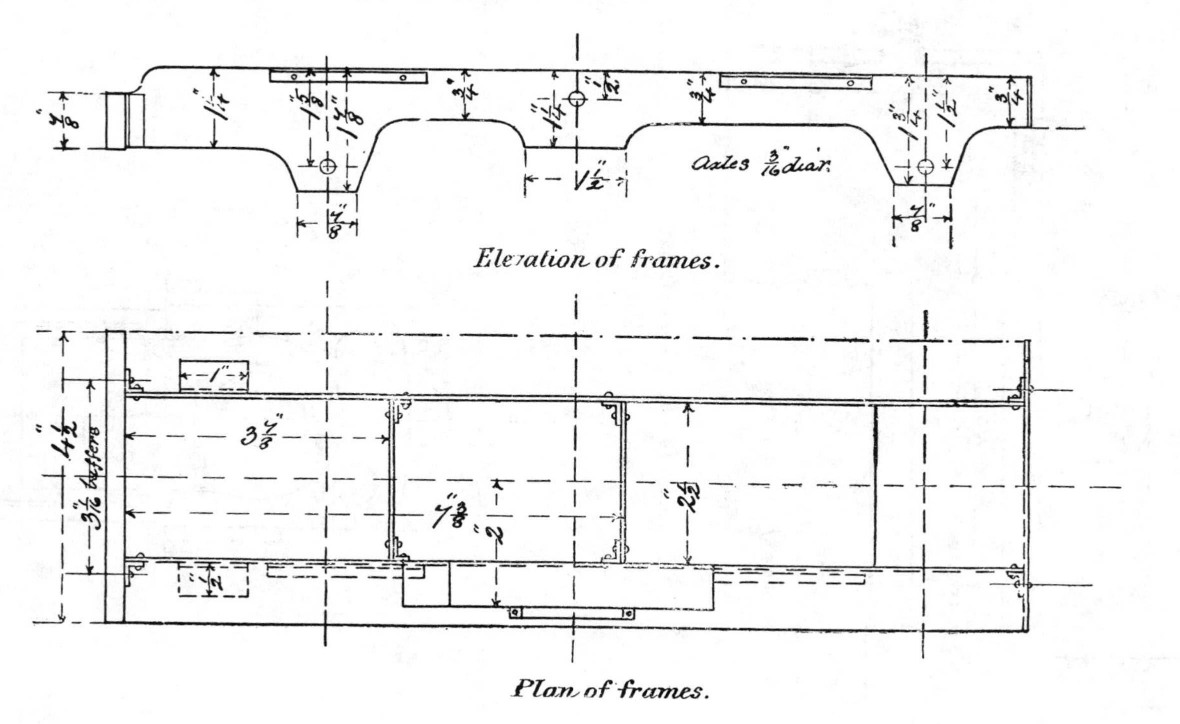
Axles 3/16" diar.
Elevation of frames.
3 1/16" buffers
Plan of frames.

SINGLE LOCOMOTIVE.—SHEET NO. 13.

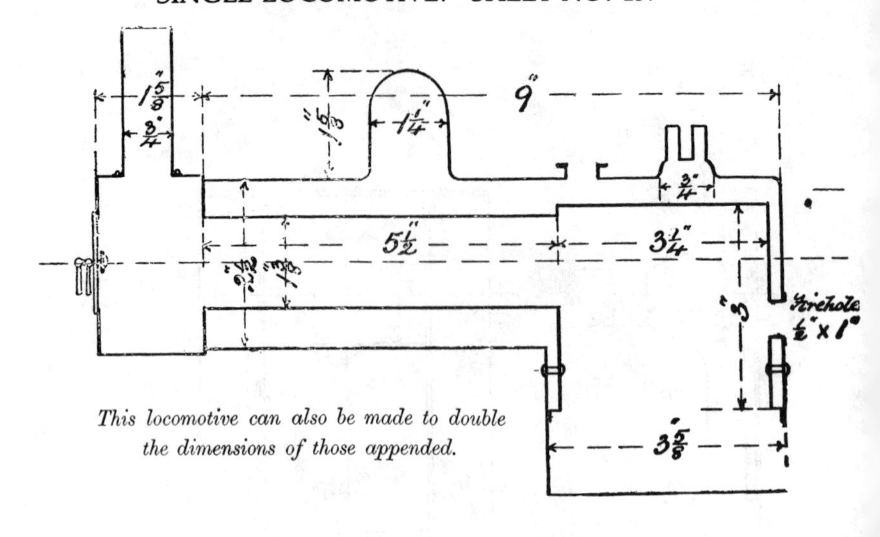

This locomotive can also be made to double the dimensions of those appended.

Longitudinal Section of Boiler.

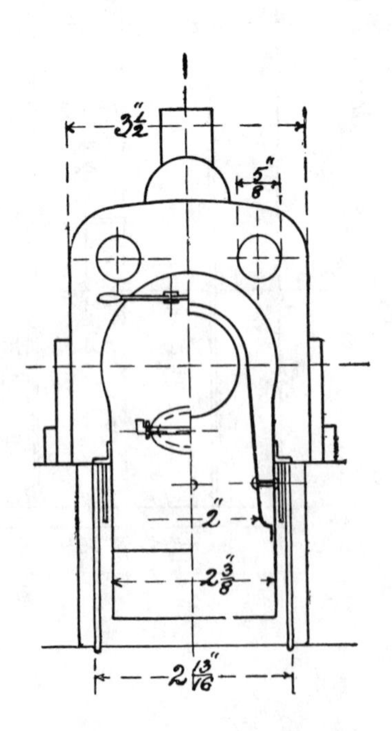

Half end elevation. Half transverse Section through Fire Box.

SIX-COUPLED TANK LOCOMOTIVE.—SHEET NO. 14.

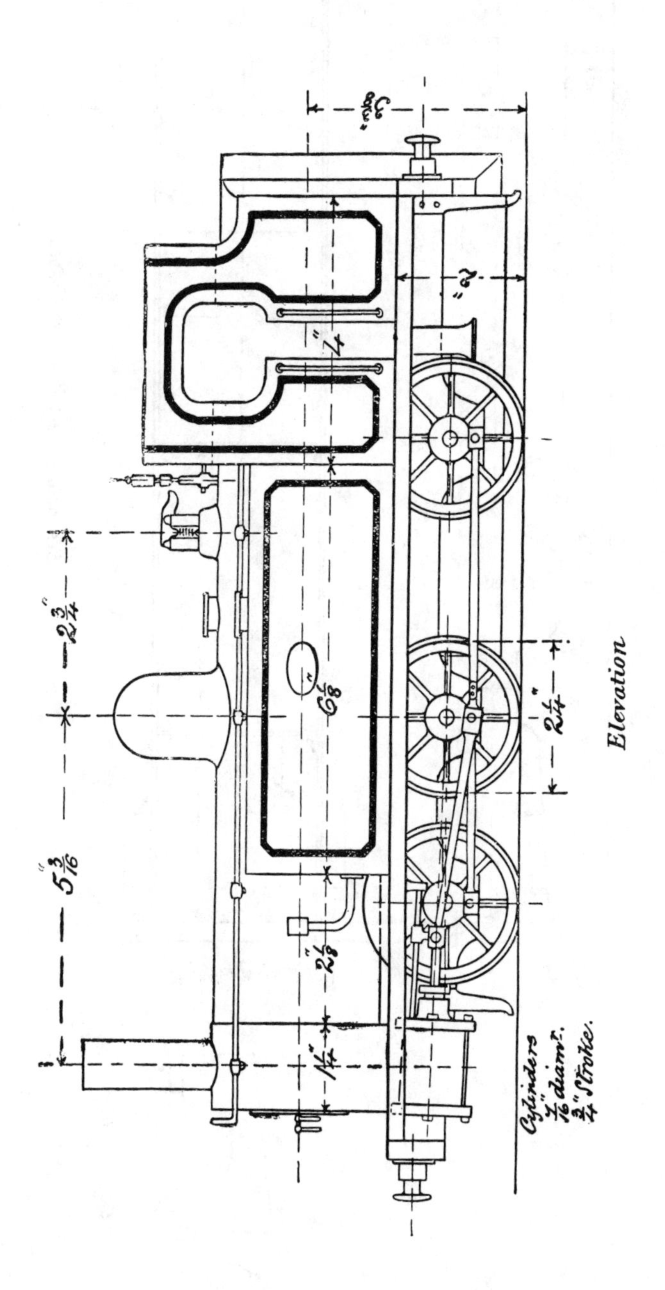

SIX-COUPLED TANK LOCOMOTIVE.—SHEET NO. 14.

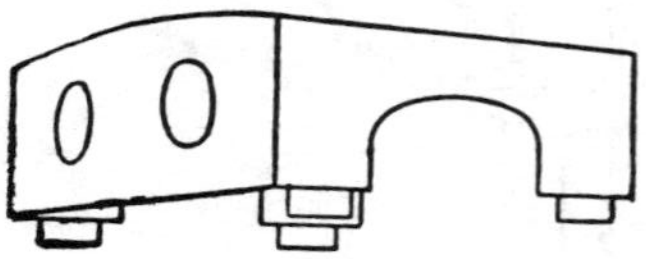

Roof of Engine Cab.

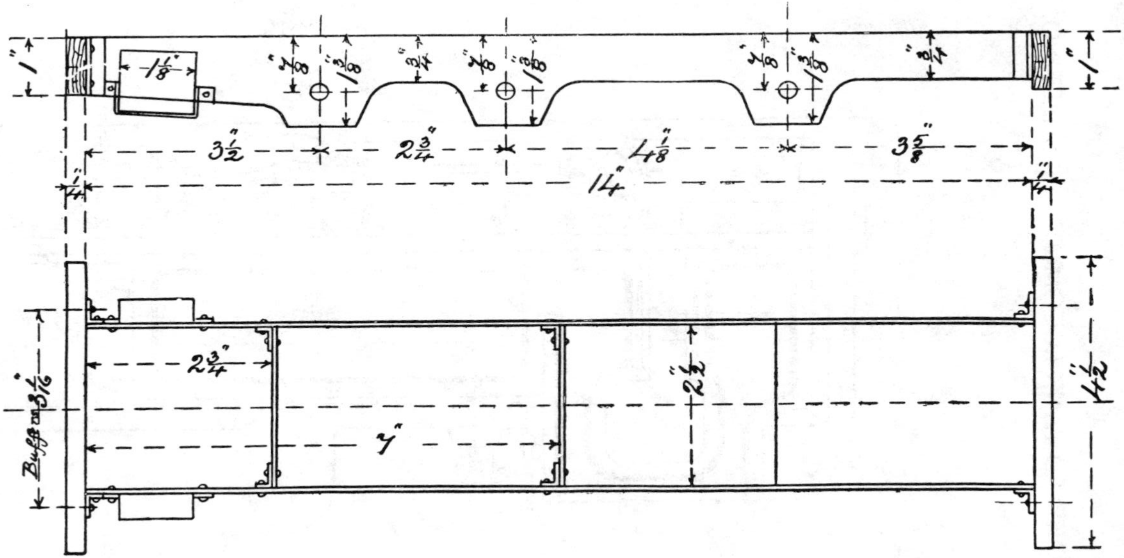

Elevation and Plan of frames.

…LOCOMOTIVE.—SHEET NO. 14.

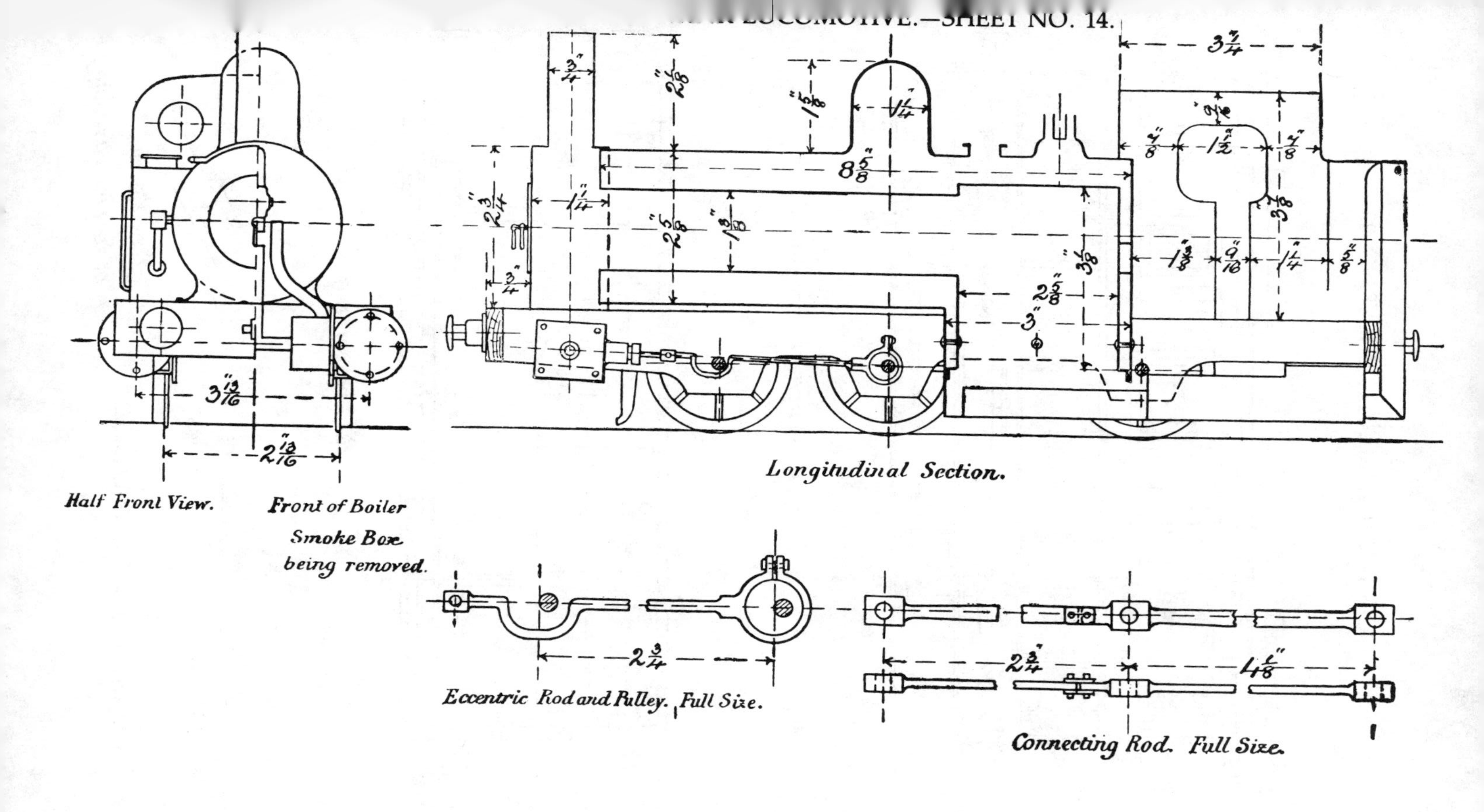

RAILWAY CARRIAGE.—SHEET NO. 15.

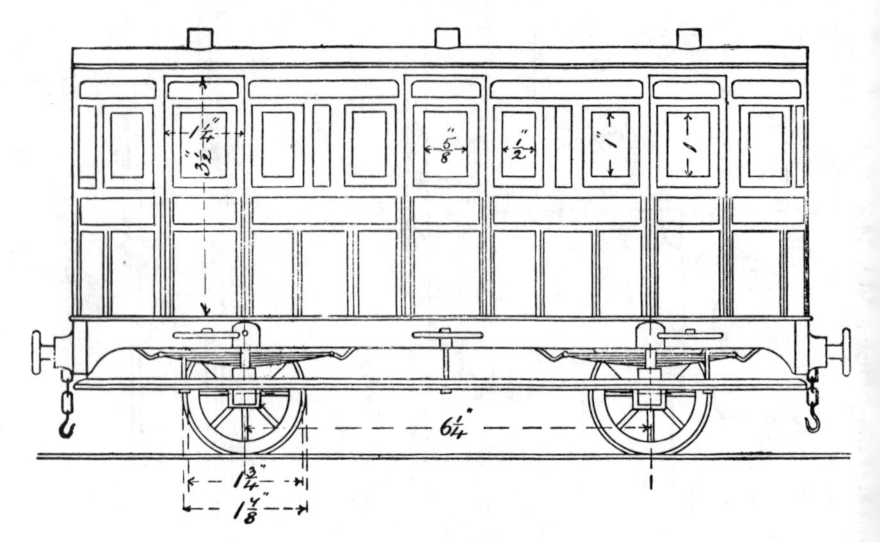

Elevation of carriage.

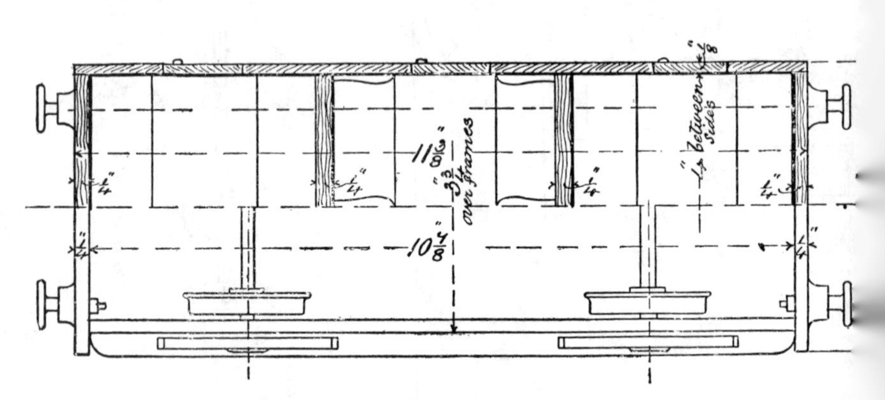

Plan of carriage.

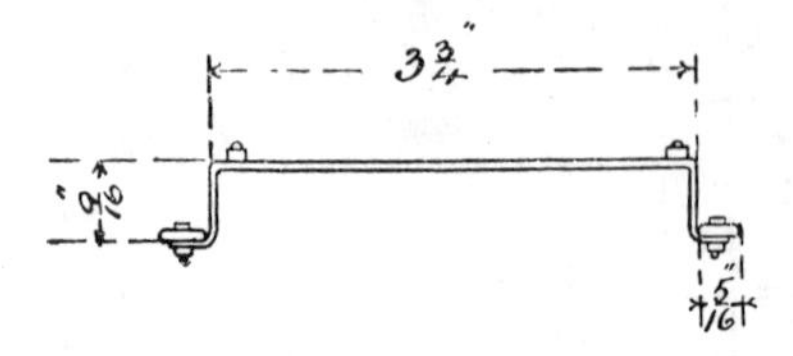

End view of foot boards & hanger.

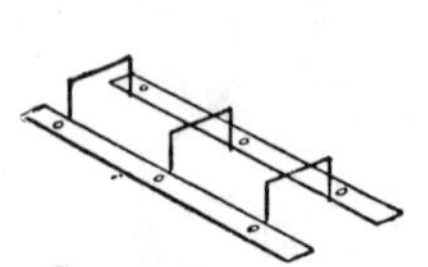

Pers sketch of foot boards & hang

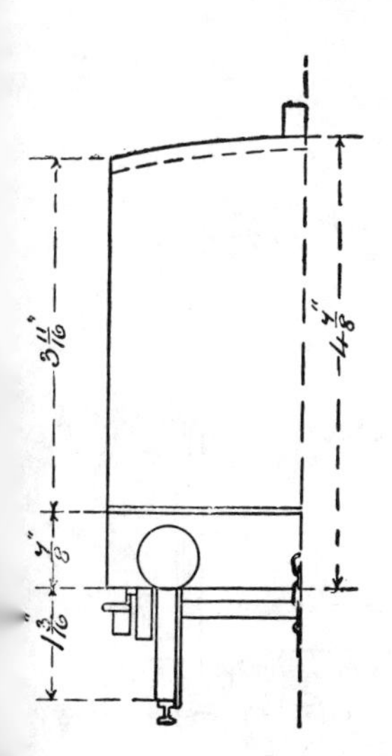

Half end elevation.

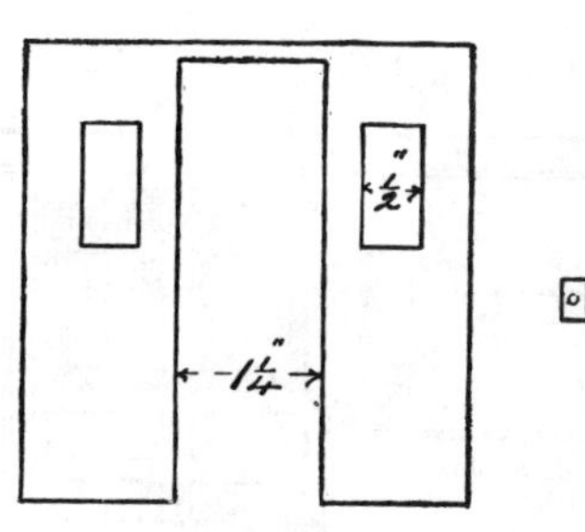

One of the sides of a compartment with doorway.

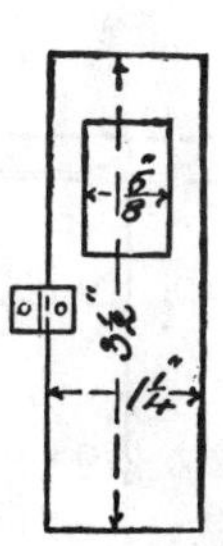

A door with the hinge attached

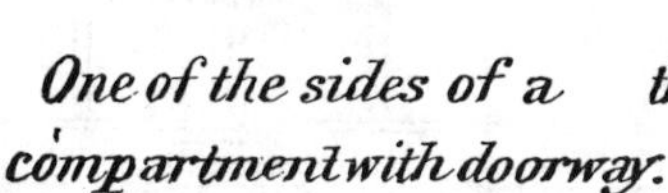

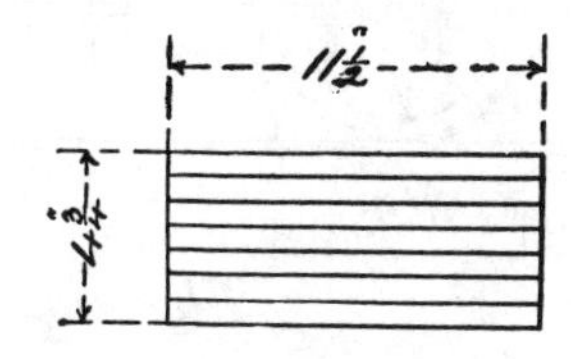

$\frac{7}{8}$ Full size..

The roof of the carriage the longitudinal lines show the saw cuts, made in the wood before steaming it, to get it to bend.

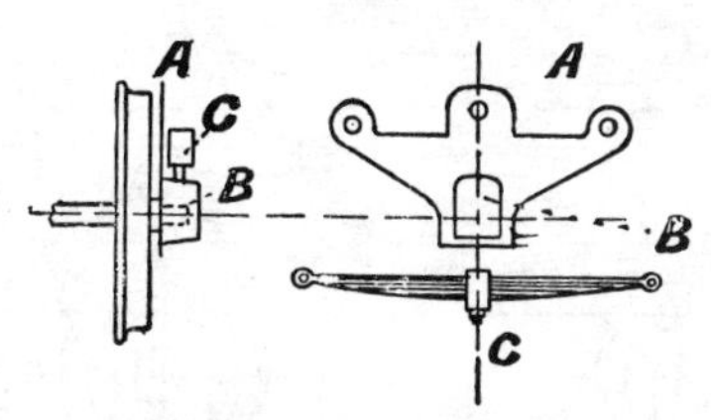

A Sheet brass bearing.

B Axle box soldered outside.

C Spring composed of plates of sheet brass surrounded by a strap.

RAILWAY VAN.—SHEET NO. 16.

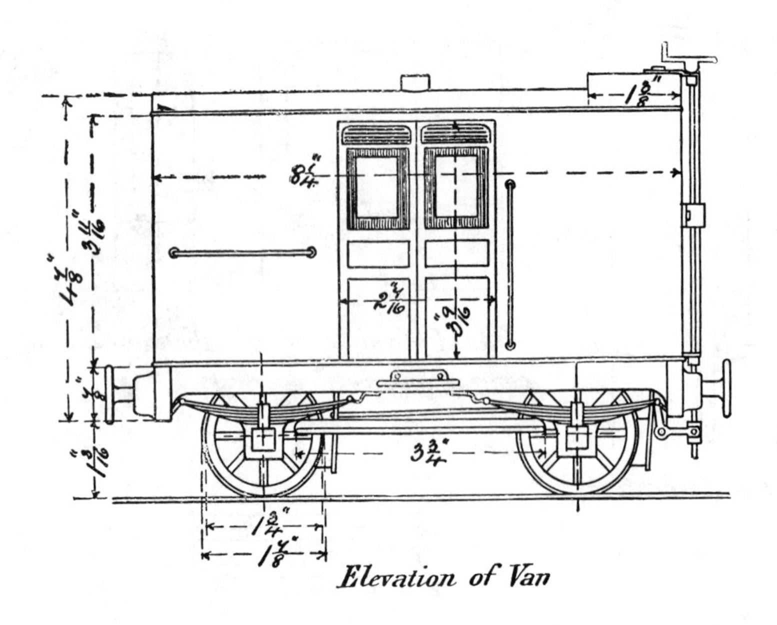

Elevation of Van

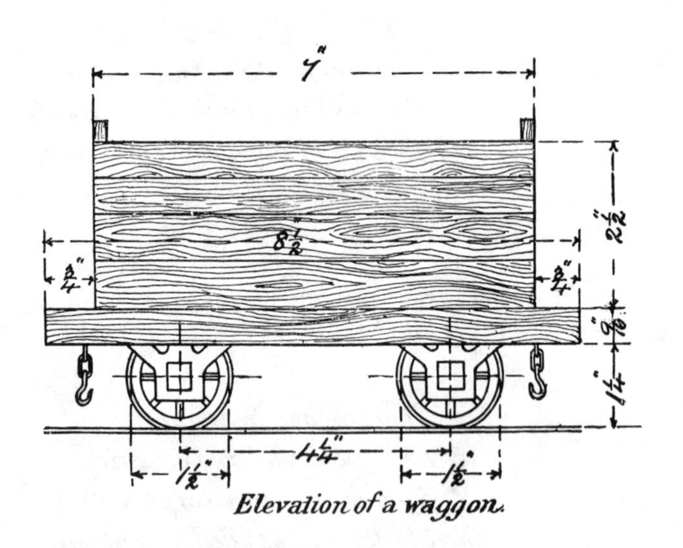

Elevation of a waggon.

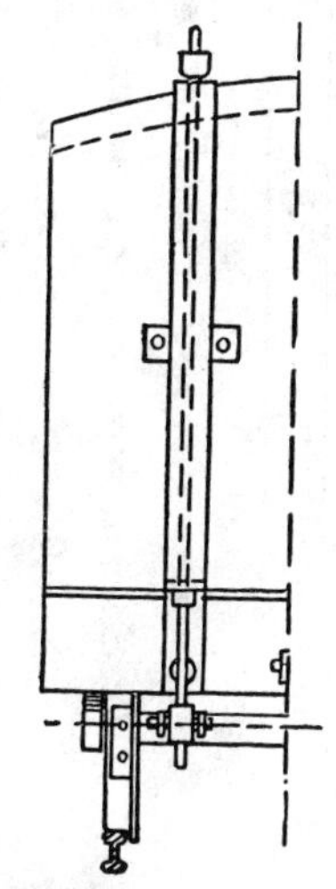

Half end elevation Showing attachment of brake, lever & rod.

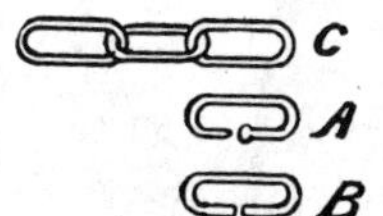

Full size.

Chains for coupling the carriages together.

A. Pin bent oval.

B Pin minus head & point.

C. The complete links attached to form a chain the ends of each pin being soldered after the links are fastened together.

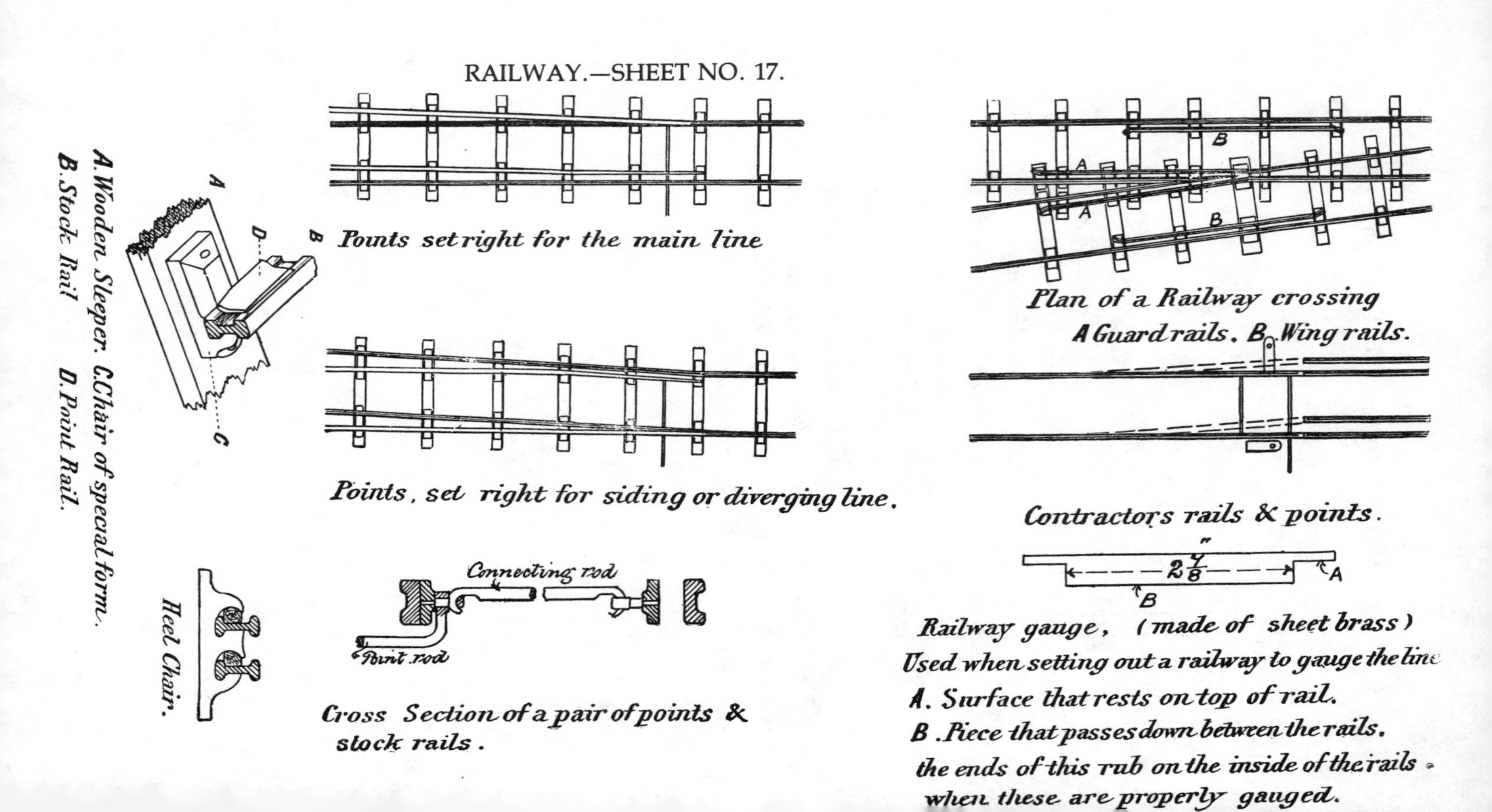
RAILWAY.—SHEET NO. 17.
Points set right for the main line
Points, set right for siding or diverging line.
Connecting rod
Point rod
Cross Section of a pair of points & stock rails.
A
B
C
D
A. Wooden Sleeper. C. Chair of special form.
B. Stock Rail D. Point Rail.
Heel Chair.
A
B
Plan of a Railway crossing
A Guard rails. B. Wing rails.
Contractors rails & points.
2 7/8"
Railway gauge, (made of sheet brass)
Used when setting out a railway to gauge the line
A. Surface that rests on top of rail.
B. Piece that passes down between the rails.
the ends of this rub on the inside of the rails when these are properly gauged.

RAILWAY SIGNAL.—SHEET NO. 18.

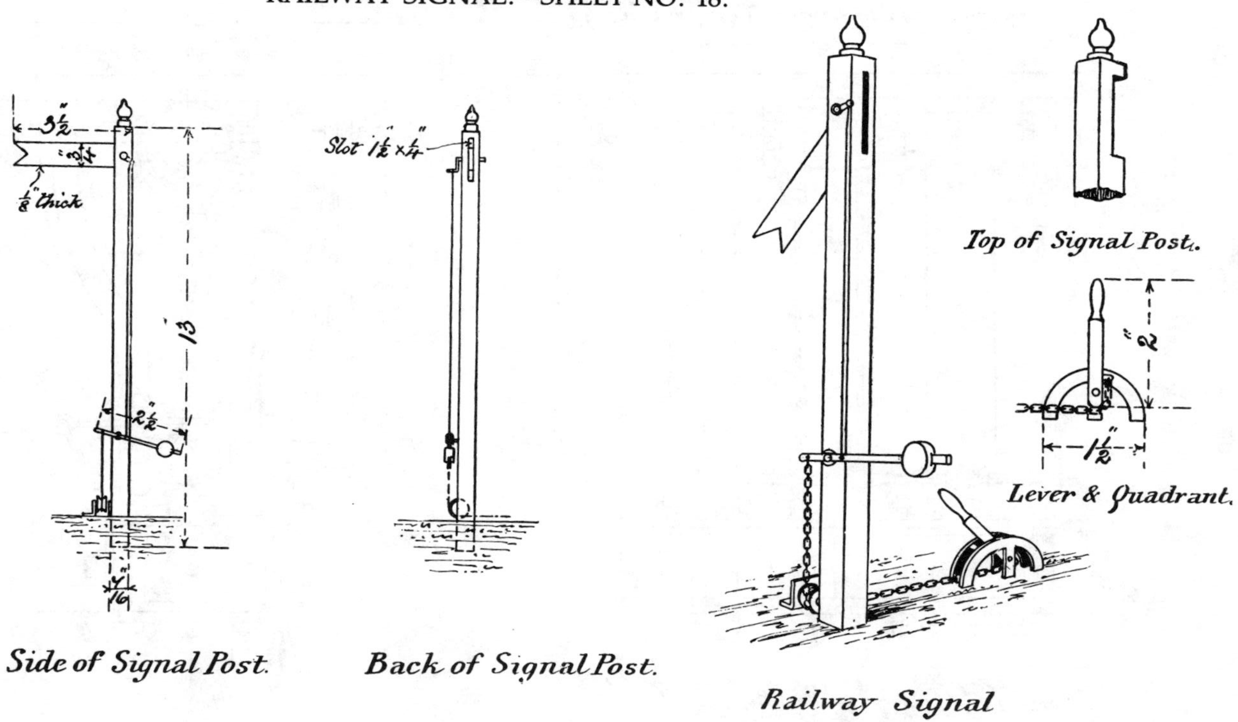

MARINE ENGINES.—SHEET NO. 19.

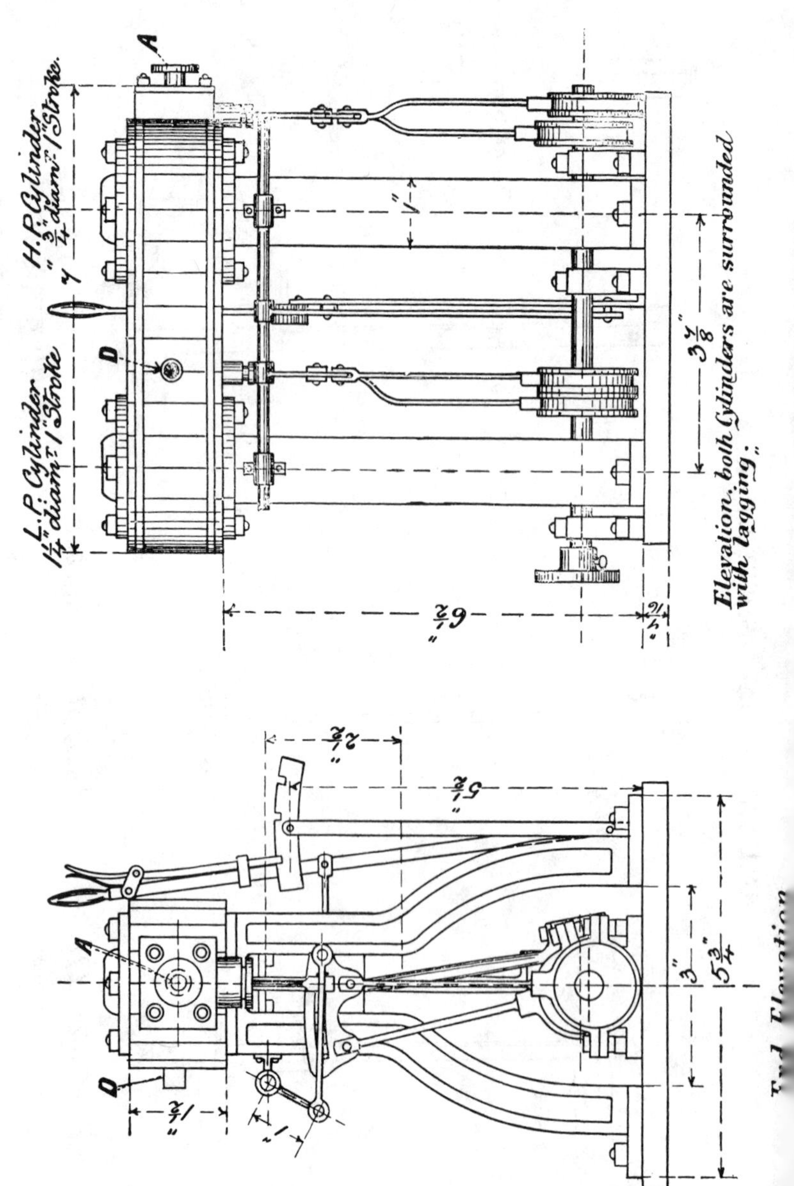

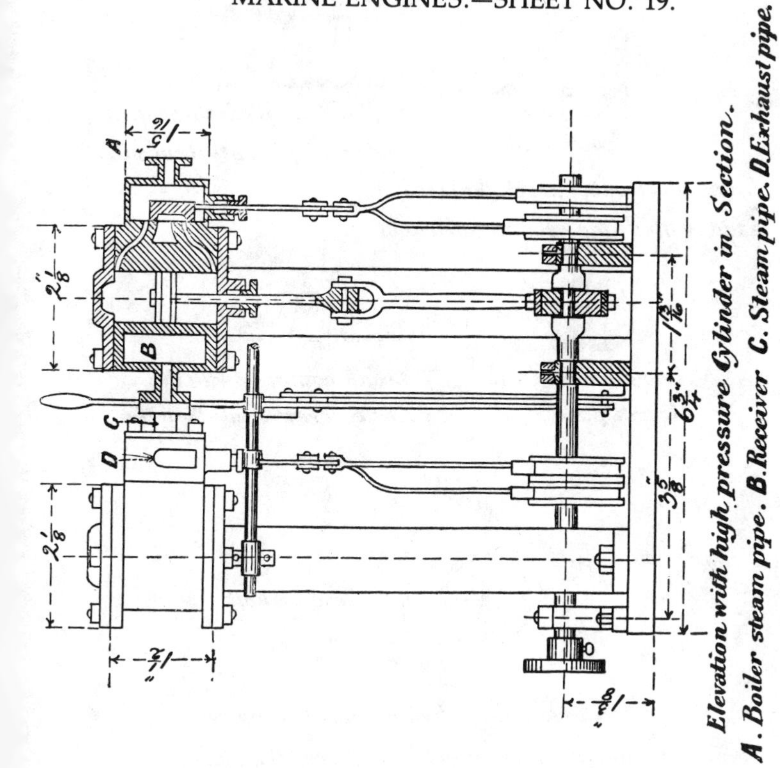

Elevation with high pressure Cylinder in Section.

A. Boiler steam pipe. B. Receiver C. Steam pipe. D. Exhaust pipe.

Plan of Engines.

Lagging is shown surrounding both Cylinders.

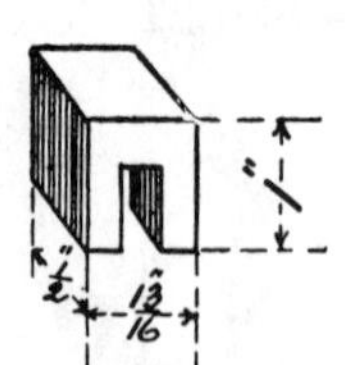

Brass block out of which the Crank is to be made, the webs are slotted out.

The Block fitted into shape, but Crank pin not yet cut out.

Crank finished and Crank pin turned.

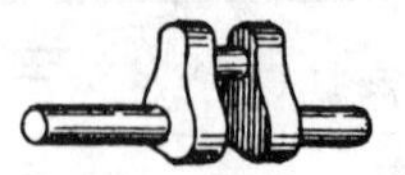

Crank complete with the Axle screwed into the webs.

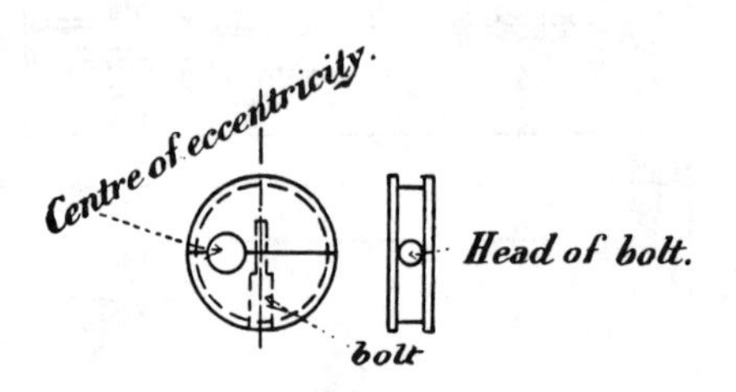

Eccentric pulley divided into two halves for the purpose of getting it over the Axle.

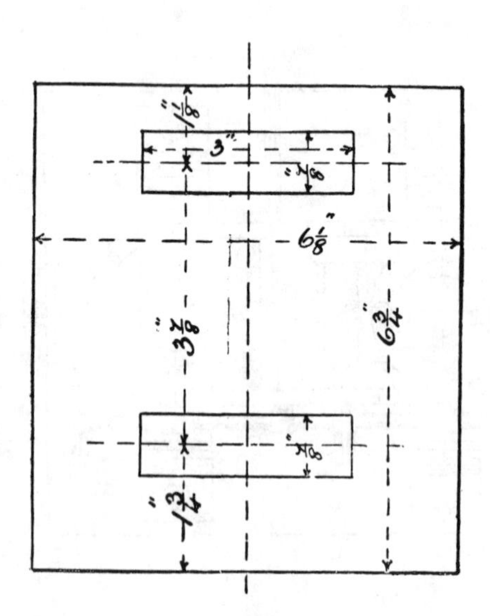

Plan of bed plate.

COMPOUND TANDEM ENGINE.—SHEET NO. 20.

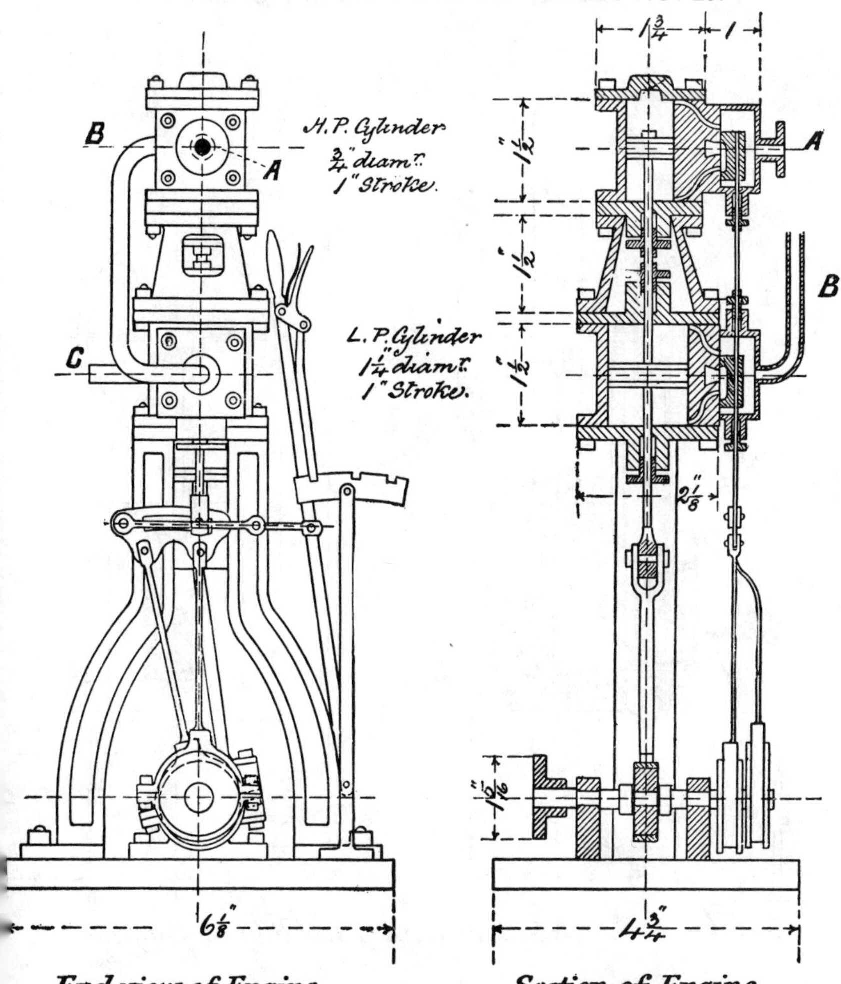

End view of Engine. *Section of Engine.*

A.* Boiler** *Steam pipe* ***B. *Steam pipe from high to low pressure Cylinder,*

C. *Exhaust Steam pipe.*

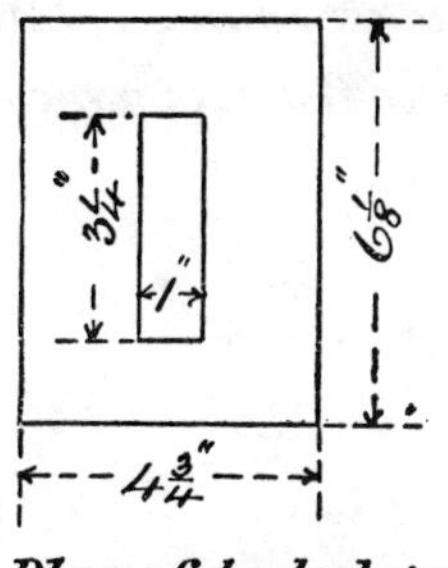

Plan of bed plate.

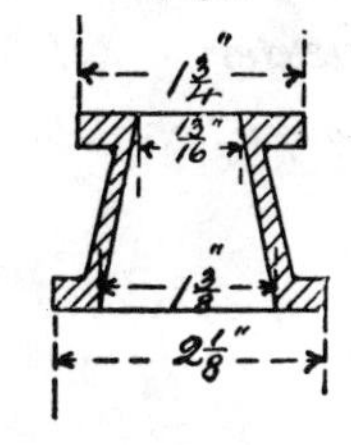

Section & perspective view of distance piece between the cylinders.

HOT-AIR ENGINE.—SHEET NO. 21.

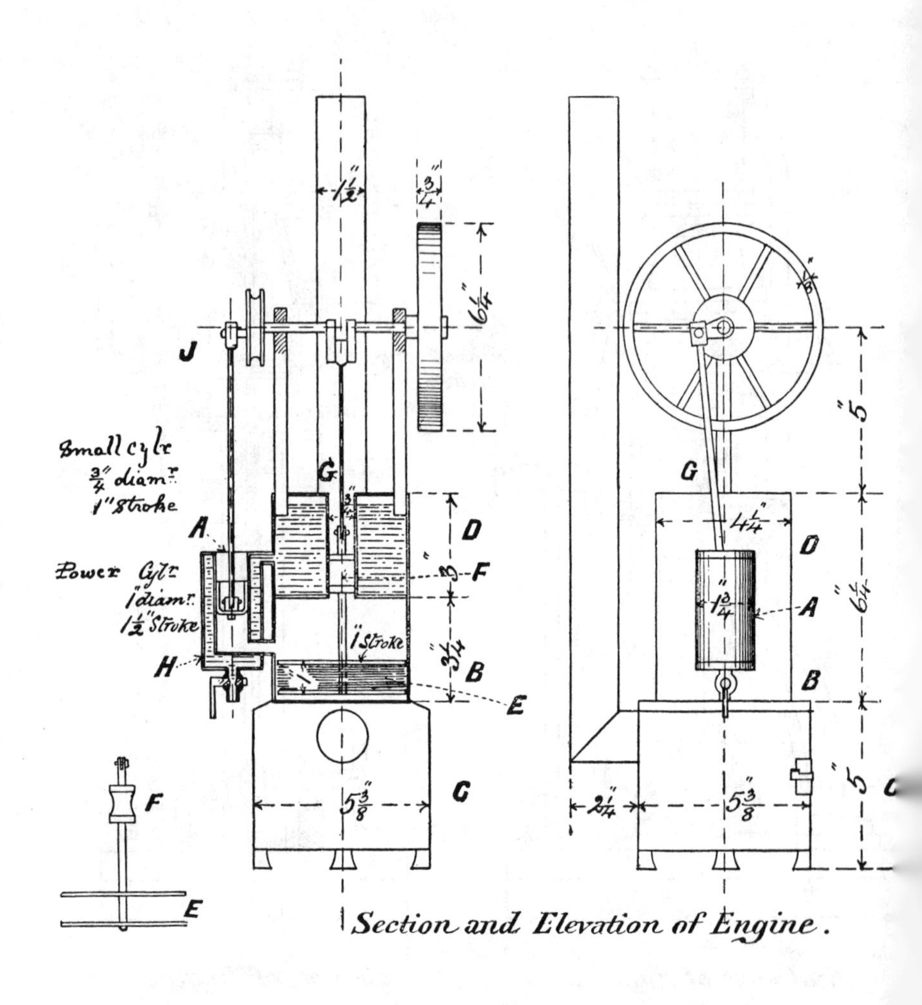

Section and Elevation of Engine.

A. Working Cylinder.

B. Heater.

C. Fire Box.

D. Water Jacket over heater.

E. Compression Piston.

F. Small Piston.

G. Connecting rod to centre cran

H. Water Jacket round Cylinde

J. Power Crank to which power Piston is attachea

HOT-AIR ENGINE.—SHEET NO. 21.

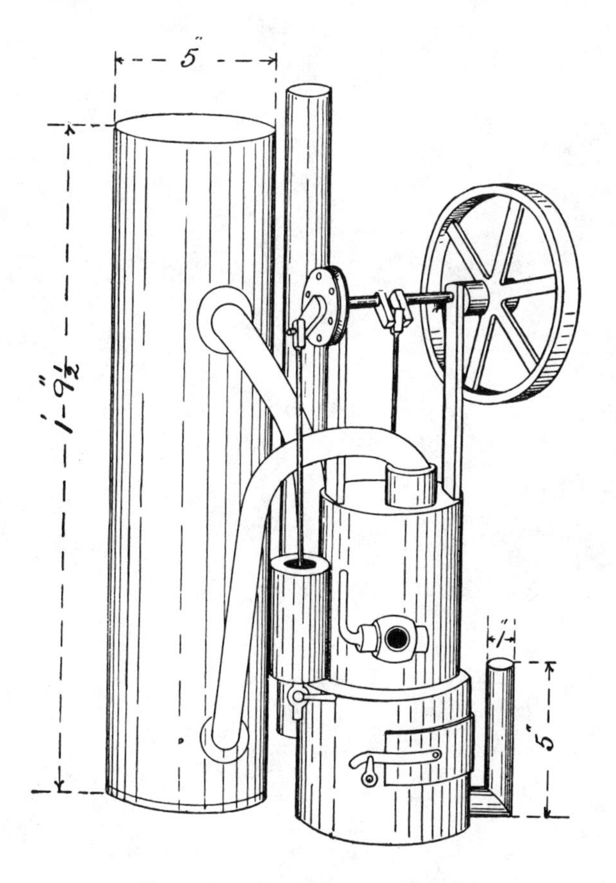

Perspective view of Engine.